'TIL THE COWS COME HOME

The Almost True Story of a Bicentennial Farm in Epping, New Hampshire

---

Carol Harvey Clapp

Illustrations and Limericks
By
David Lyman Clapp

'TIL THE COWS COME HOME –
The Almost True Story of a Bicentennial Farm
In Epping, New Hampshire

at P.O. Box 578 – Epping NH 03042 – USA or
33 Morton Street – Riverton 9822 – New Zealand
or codaclapp@gmail.com.

Cover: USDA Whole Herd Termination Program of 1986-87. Phyllis in horse barn as 20 cows are loaded for long trip to Jordan, donated to Heifer Project International. Other 125 were slaughtered.
Back Cover: Me in China near Terraces of Longji.

ISBN 978-0-615-38267-8

Printed in USA by 48HrBooks
(www.48HrBooks.com)

Written for

The Grandchildren who will someday ask: “What happened?”

FORWARD

by Dudley Laufman

A SWALE OF GREEN ON RED OAK HILL

Back in the late 1940s, Jonathan and Betty Quimby held square dances in the kitchen of their 200 year old farmhouse in Fremont, New Hampshire. These parties were called "Kitchen Junkets." Bucky (Jonathan) played the fiddle. Ruth Chandler or Mrs. Sheldon played the piano while Betty called out the changes to the Virginia Reel. The house smelled of woodsmoke and faintly of cow, as did all of us who worked in the barn or dairy. I was proud to wear that odor and to talk New Hampshire.

The neighbors came to be at these shindigs. Dan Harvey and his wife Louise from their dairy farm on Red Oak Hill in Epping were always there, with their New Hampshire accents. Dan was too tall to dance. The room had a very low ceiling and his head didn't clear the rafters. He would retire to the dining room to play cards with Bill Roberts, Raymond postmaster, and the Reverend Edgar Chandler.

Phil Peterson, Louise's brother, had a dairy farm on the Fremont, Epping line. They had a hip roof barn. Looked like those barns in Wisconsin. From North Road you could look down across what seemed like miles of meadows and fields to their place in the distance. On June 9, 1953, late in the afternoon, Phil saw a red tornado churning through the woods just west of North Road, on its way to Exeter. At first he thought it was a large flock of birds whirling around. That was the same day as the killer twister that ravaged Worcester.

I was walking one day (half a century later) on the abandoned railroad bed (Worcester-Portland RR). Philip's Holstein bull stood right in the middle of the way and would not let me by.

Recently Jacqueline and I did a program at the Epping Historical Society…some fiddle chunes, poetry, stories. I made sure that they invited Dan Harvey, because I knew he would help us along with a story or two. Sure enough. "You know that marsh hay they cut on those tidal fields, what they feed to horses, cows won't touch it. Well the same is true of that meadow grass we have here, cows would sooner die than eat it. And we have lots of it from them low lying fields of ours. Cows just push it aside, maybe pick through it a bit try to find a sweet morsel. So we figured out a way. Had our neighbor help us and we used his truck to bring it in. Cows see a strange truck coming in with the meadow grass, they think we had to buy it, so they'd eat it out of guilt."

Dan Harvey is a character, no two ways about it. It is the way he is. I do not think he works very hard at it. Over the years I have only known him by way of stories from the Orzechowski family and from Frank Ferrel, and from occasionally seeing and talking with him at various events. Carol's story opens up another side to the Harvey family, some of which might be called dirty laundry. But it all is what makes us human with foibles and weaknesses. We can still be characters. And it is important that we carry that faint perfume of cow wherever we go.

In many non-ethnic New England communities, occupations skip generations. A farmer's kids head for town, their kids come back to the land. My grandfather was a millwright/farmer, and he played the fiddle. My father was a Boston businessman. I took to the woods and play fiddle. My kids went the urban route. But my grandkids are showing leanings towards bluecollar rural and music. So, in spite of cows coming and going, living off the land is still an option for the Harveys in Epping. Phyllis has corralled some black and whites and makes cheese, Analesa grows veggies, Sharon lives in the woods close by, Amos and Otis are in the neighborhood along with Danny who works the old stumps with his dad. The point is, they are sticking around on that swale of green.

Dudley Laufman
Canterbury
New Hampshire
October 2009

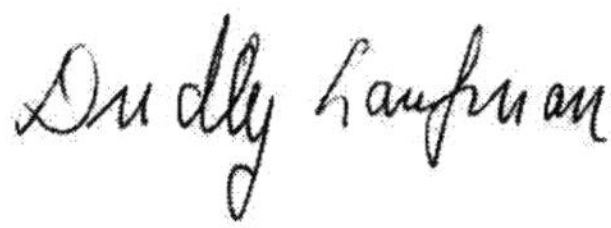

Epping Historical Society, Summer 2008

'TIL THE COWS COME HOME

The Almost True Story of a Bicentennial Farm
In Epping, New Hampshire

PREFACE

Two years ago, I began writing this story about growing up on a dairy farm. I covered the usual stuff like fond memories of cows and bewilderment as the oldest child in a big family, lamenting the politics and societal pressures on our lifestyle in New Hampshire's fast-growing (Rockingham) county.

During this exercise in memoir, I have felt great sorrow acknowledging that our family farm is a gonner—not as much fun as it used to be. While Applehurst Farm remains physically intact, its agricultural purpose has been gutted, its unified spirit broken.

My sons are deprived (or spared) the opportunity to work together on the farm. They will not share responsibility for morning barn chores. They will neither milk cows nor raise calves to replenish the herd. They will not worry about silage trucks getting stuck in muddy fields or become despondent over property taxes. My grandchildren will know nothing of driving tractors with power-take-offs and leaky hydraulics. They will not till rocky earth with plows and harrows. Neither will they raise calves, pigs and colts or toss bales up from one to another until the hay mows are stacked 20 high.

In these modern times, PTOs, small bales, hay mows and tossing make no sense.

Animal husbandry as I experienced it has no place in our modern economy. It is old fashioned, not profitable, too small scale and best left to hobby farmers.

Toward the end of journaling, I was beginning to wonder if The Harveys are bad enough to write about. Compared with other families I've read about, we're not very exciting – no incest or abuse, no henchmen or bank robbers, drug addiction or abject poverty. Heck! If it weren't for a neighbor, we wouldn't even have murder and sexual exploit.

In The Beginning ...

Just as I was indulging in literary doubt – and through no fault of my own — two of my farming uncles died and a plane crashed on the hill. Even better, my five sisters began practicing tough love by chasing each other around with brooms. These deaths, accidents, and feisty behaviors spiced things up a bit, providing material to show how upsetting it is to lose family farms in general, and ours in particular.

"We are NOT fighting!" snap The Sisters. Just straightening out a few things, showing concern for one another's wellbeing and getting rid of the cows. "We HAVE to be tough. Things NEED to change! ... It's just business, for Heav'n-sake!"

Without doubt, today's Applehurst Farm is looking good without a dairy herd. The barn has a fresh coat of red paint and, (well, actually, it's vinyl siding) and finally – after many attempts, all unappreciated — our father's junkyard is nearly cleaned up – (not that everyone likes it 'cleaned up'.) The Fernald Family leases our cropland for hay and corn to feed their 200 cows in Nottingham. They are still farming.

My youngest sister Erica cremates dogs and cats for a living. She does a premium job of it and it seems to pay. She and her husband Steve operate three incinerators that roar and drive our sister Beth nuts. But nobody pays heed to Beth's complaint because Cousin Margaret heard noises, too, and it didn't amount to anything. Erica has generously remodeled our farmhouse and rents it to friends. Before this, she financed a new house across the road from where we all grew up.

Thanks to Erica, our parents now enjoy a comfortable single-story home where they can monitor farm activity from their deck. My youngest sibling pours thousands of dollars into farm improvements — ever larger lawnmowers and vacuum cleaners. New golf carts and rototillers for Mom. She has restored the barns and sells ice cream where another sister (Phyllis)'s bulk milk tank used to be.

I know! It's already too many Harvey girls to keep track of. Four, thus far: Beth and Erica aka "The Babies" (because they were born six years after the rest of us). Erica the generous businesswoman; Beth who is divorced with two kids and lives across the road from Erica; Phyllis The Cow Girl; and I'm Carol The Oldest at 62. We'll get to Susie and Cindy later.

Right now, because of the ice cream stand, Dad (89) is rattling the neighborhood with his excavator. He is installing a septic tank behind the barn. At last! The Harveys are catching up with my mother's side of the family — The Petersons — by putting a toilet in the barn! — except that Grampy Peterson's 1930s barn loo skipped the septic and provided a long drop into the manure pit. Dad's crashing and thrashing remind me of when, 50 years ago, he evaded the planning board by secretly digging graves to expand our cemetery behind the barn.

Dad is a lifelong digger. It's so strange now to see this puny septic system overshadowed by a mountain of cow manure beside an even more incongruent federally-funded concrete manure storage pit. It is huge. But empty.

Strikes me funny. All this capacity for waste management. No threat of polluting downtown Epping. Narry a turd will escape Red Oak Hill. Not to mention Dad's odd effort to comply with, rather than ignore, the building code. My, how things have changed!

They call the ice cream stand "The Milk Room" ... which it was until the dairy was shut down last year.

I'm not sure about milk in soft-serve concoction delivered to the farm in boxes each week. It's certainly not milk that Cow Girl Phyllis can't get rid of at her own place ('Edgerlys'') on the other side of the Night Pasture. That's where she retreated with the young heifers, now calving. When her surplus milk goes sour, Dad takes it to Fremont for Uncle Philip's pigs.

But Philip isn't there to enjoy the pigs' lusty squeals and gulps, since he's dead. Died last summer, just after Uncle Bill.

Joe Bodge makes 30 flavors of ice cream from scratch, downtown near Star Speedway. He heard that our farm is going into competition with him. Joe is not enthusiastic about soft serve. He says it's not real ice cream. We suggested that he sell his real stuff to The Milk Room.

"No way in hell that's gonna happen," he declared, asking: "Can you feel the love?"

Good question. Where IS the love?

Joe knows all about Phyllis being evicted and has helped her – loaned her a tractor, baled some hay, listened to her story. He tends to sympathize with her and the cows. And, while writing about growing up on the dairy farm, I, too, cannot deny my sorrow over having it end this way. This book is more about the lost dairy farm than about my part in its demise.

So, yes. Applehurst Farm is looking better than ever. Mothers and children are laughing and licking cones. Soon there will be doughnuts from the same recipe Mom used for after-church 'Fellowship Hour'. Police are already circling, settling for ice cream until the doughnuts are fried. We used to avoid cops, but now they're supposed to be our best buddies.

At the adjacent vegetable stand, Mom's tomatoes, string beans, and cut flowers are selling like hotcakes. Remarkably, in just two days, Fernald Farm swept our fields clean of second-crop hay. We never owned such good harvesting equipment. I inhale as the crop is trucked away. David Fernald's friendly wave and the sweet fresh-mown hay aroma prompt so many memories of growing up on the farm.

Nieces Christa and Abbie's new little white house is cute beside the farm pond. It is similar to 'The Little

House' of my childhood when on muggy summer nights we six oldest kids mumbled stories through air holes in itchy blankets, hiding from mosquitoes.

While walking through my mother's garden, I see Fernalds' hilltop corn crop reflected in the still pond water. Mom's sunflowers and brilliant zinnias adorn her vegetable garden as they have since my childhood. Cheerful farmhouse tenants come and go as if they own the place. They are laughing. I've heard people joke that the tenants are the kids our parents never had.

Harvey Farm is bucolic, solvent and sweet, now that those darned cows aren't draining the finances and crapping everywhere. So calm and pretty.

Yet I feel an unexplained disturbing sense of bereavement for our lost lifestyle.

"Don't cry over spilt milk!" they say.

Still, I miss the cows and wish we could all come home.

Cider apples headed for press from ancient Baldwin orchard that attracts UNH field trips because it's so unusually old. Dad, Matthew, Phyllis. (hired man) Alfred Carr on load

In The Beginning …

SPILT MILK

Phyllis should have said: "No, Thanks!" when Uncle Philip offered his cows in 1994. Instead, my sister accepted the 24 Holsteins and began putting the dairy back together, buying a new bulk tank and requesting the return of equipment that had been loaned in 1987 when Applehurst Dairy was terminated the first time.

If we had known in 1994 what we've since learned, certainly none of us — maybe not even Phyllis — would have continued the Brookvale-Applehurst bloodline. Then again, my stubborn sister seldom chooses the easy path and how else would we know what to expect when we bury the folks and each other behind the barn?

Our Harvey situation isn't so different from other American farms where eight adult offspring refuse to leave home and all look alike.

People downtown sometimes guess our identities, but usually don't know who we are — other than Harveys who aren't too bad. We are not snappy dressers. Neither are we very fat, yet no one has ever called us 'skinny'. It pains people to look at our eyes because we all squint so much. Our sensitive Scandinavian skin turns red when exposed to the weather which is where we spend most of our time. I've been asked if my nose is red because I drink. Nope! I don't drink.

I look like my five sisters, but mostly like my brother Albert who has a mono-brow and sparse eyelashes. My sisters look like each other: Beth resembles Cindy; Erica resembles Susie; Phyllis looks like Matthew. Two redheads: Albert and Erica. One notable blonde: Susie. People say we all sound the same on the phone with our cracking tones.

Phyllis is distinguished by having the biggest cowlick. Her hair swirls in the middle of her forehead. Playing as kids, we all laughed so hard when a Newmarket farmer, Gerard Hamel, shouted: "Hey Bushy! Go grab my cigarettes off the dash of the truck, will ya?" And everyone knew he meant Phyllis.

Did I mention there are eight of us? Dad still says: "I have eight kids and they're all boys — except six." I'm not sure he's as sure about the number of grandchildren, let alone great-grandchildren.

Plus, we have fifty other look-alike relatives all living within walking distance in and around hometown Epping. With a genetic cluster like that, you'd think we are hunkered down in a forgotten pocket of Appalachia, instead of draped across the town's two highest hills in southern New Hampshire within spittin' distance of Boston.

Seems like some of us should've crossed the line into Massachusetts and gone there, to Boston. Or Lawrence or Lowell. Maybe Methuen or Haverhill, at least. Economic opportunity is that close. Isn't that what they did in the old days? Anyone with a bit of gumption copped an attitude and ran off to the city to work in the textile mills and now the high-tech industries.

Or they went to war. The Revolutionary War, the Civil War, the World Wars, Korean War, Vietnam War, Gulf War, Iraq War ... ANY war. We've had a few. But Nope! Not the Harveys. My father, Daniel Webster Harvey, puts on a pretty good show telling about how he wasn't allowed to enlist for World War II, as if he would ever follow orders.

"They wouldn't let me be a hero and come back all decorated with medals," blubbering, "Boo hoo!" claiming he had to stay home to grow potatoes as the only son in a farming family. He explains that my grandfather bought him a new potato digger and a Piper Cub. "He told me to join the Civil Air Patrol and guard Hampton if it would make me feel better."

So that's what he did. He stayed home to watch over us and has done so ever since. He worked all his life to be an independent New England character and has succeeded brilliantly. Although often outspoken and sometimes irritating while lecturing us about politics, "the powers that be," and the responsibility of owning land, Dad is generally respected, a lovable brat, basically benign and cherished by family and friends around Epping and throughout Rockingham County.

The police chief told me just yesterday: "I do enjoy your father. He's an Epping icon. That's for sure! I could listen to him all day. ... Come to think of it. I HAVE!" We laughed comfortably within the trust of our lifelong familiarity.

In The Beginning …

Between you and me, I think Dad just didn't want to leave the farm. Sticking right there on Red Oak Hill is what we Harveys do best. In 1976, we even received a prize for it! National and state recognition for being a "Bicentennial Farm" — same family on same farm for over 200 years — rewarded for not budging.

You see there's nothing new about my story. Like most other New England farmers, we began going out of business as soon as we tried going into it. Market conditions were never favorable – especially when long-distance freight and technology started competition with New York State and the Midwest. We did better when we were self-sufficient. Used to be that each household kept a cow, even downtown, and grew a large garden. There was no food industry as we now know it, no bulk processors and distributors. Everyone grew and made their own everything and bartered with the neighbors when they had to. There wasn't so much paper money to worry about, back then.

Progress, marketing and economic sense came later.

Not surprisingly, my mother's family did the same thing on a farm just seven miles from Red Oak Hill. My Peterson grandparents (Albert and Eunice) didn't start out producing milk for sale. They waited until there was an off-farm demand, around 1920, then began milking 20 cows by hand, packed 10-gallon jugs in ice, and hauled them to Epping Depot to go by train to H.P. Hood in Manchester.

Concurrently, so closeby, my Harvey grandparents grew apples for export and for local consumption, fresh or to exchange hard cider for labor.

Both farms, Petersons' 'Brookvale' in Fremont and Harveys' 'Applehurst' in Epping, were passed down from my great-grandparents who remained on the farms until they died, cared for by my two sets of grandparents just a town line apart.

Crest of Red Oak Hill north slope hayfield was cut with sickle mower; more recently with mower-conditioner. Now planted to corn for silage harvested by Fernald Farm of Nottingham

Harveys & Petersons chartered bus to attend 25th wedding anniversary for Uncle John & Aunt Agnes Gilbert in Surry, Maine (40 passengers, 10/12/95). L-R: Otis, Abbie, Cindy, Carol S., Louise, Ann, Roland, Alberta, Philip, Aunt Ginny, Judy and Albert Peterson, Cousin Eric Schreiber. Carol Schreiber drove bus. Two years later, Mom and Aunt Ann drove to Osage, Iowa to celebrate Aunt Agnes's 75th birthday.

Wedding of Albert William (1895-1955) and Eunice Martin Cole Peterson (1902-1996) Brookvale Farm, Fremont, 6/11/24. L-R: Henry E. Cole, Caroline M. Higgins Cole, Elton Moore (hired man), Maude Muriel Cole, Albert W. Peterson, Peter Peterson, Signe M. Peterson, Rev. Magwood. (most old pictures provided by Uncle Eddie & Aunt Ann Burtt)

In The Beginning …

Mom the Happy Chicken Farmer, Brookvale Farm, Fremont.

Signed: 'Soberly Dan '38'

"He might've amounted to something if he hadn't inherited that damn farm," said Arnold Dennett in 1971.

PART ONE

Terminated

The way we are …

1. IT'S ABOUT THE LAND

"Applehurst Farm" is better known as "Harvey Farm." It endures larger in legend than in real life. People continually refer to our 2,000 acres, though the true land area is considerably smaller, perhaps even less than 1,000 acres. But, without surveys, nobody really knows.

That said, during my lifetime, the farm DID grow to over 2,000 acres as my father indulged his love for land by buying all that was offered to him. Since then, my parents have given away and sold much of the recently acquired property. Maybe more. So who knows how much is left? Applehurst's shifting stack of deeds declares acreage in terms of "more or less." Survey engineers determine exact measurements and have prompted more than a few boundary disputes.

Dad won't give an inch during these discussions. Neighbors say: "You have 2,000 acres! Why can't you yield just this once?"

What they don't understand is that we DON'T own 2,000 acres and yielding is not something my father does. He will NOT budge. Not when it comes to land and politics. Anything else, you're welcome to it!

As lifelong members of agricultural institutions, such as The Grange and Farm Bureau, my parents feel it is their responsibility to be good 'stewards', guardians of the land. Indeed, stewardship is a concept embraced by all members of my family. It was actually learned through our youthful involvement with 4-H Clubs and other agricultural organizations. Appreciating this bias goes a long way toward explaining why my family has resisted what most people see as 'progress' ... progress involving housing subdivisions and retail development within zoned districts.

We don't like it!

Yet, each year, even without shifting, our farm has moved closer and closer to downtown Epping. Distance between the two has diminished as each open space — every pasture, hayfield, and any backyard big enough to be subdivided — has been filled with houses and condominiums. Residential development has bubbled around us, east and west, north and south, defining our farm as if it was froth on boiled milk.

Whatever our acreage, Harvey land is 90 percent covered with forest and swamps. Still, there's room for some pretty large fields, by New England standards — not that 70 acres would impress Midwestern wheat and corn growers.

My broad-shouldered hollow-chested father has sat on bulldozers and, more recently, excavators for years clearing land, forever shoving back trees along the edges of our fields. All during my growing up, he pushed to "make new land," claiming fresh territory. Now, nearly 90 years old, Dad revisits and maintains acreage he cleared during his youth.

Our land is "bony" till, full of rocks that we kids dutifully "picked" each spring. Tons of heavy dusty, soil-laden granite deposited by receding glaciers.

Well before our adolescence, Dad expected us to lug and toss rocks into his bucket tractor so he could roll them into a dump truck that he would reverse into the woods and empty into a pile.

As we got older, we filled the truck ourselves using a Bobcat loader that often tipped forward into the dump body because the bucket raised high and the rocks were so heavy. Same routine with backing the truck into the woods. We did it, like Dad, regularly getting stuck and needing to pull ourselves out with a tractor.

Even today, 50 years later, rocks are picked each spring before planting. Deep winter frost still raises them to the surface. But it's no longer our problem, since another farmer (Fernald) leases our land. I doubt that any of us Harveys are willing or able to do it now anyway. (I can hear my mother objecting: "Of COURSE we do! What's she talking about?!)

Rock picking gave us a most direct feel for our land, its familiar unremarkable soil – so unlike rich soil enjoyed by farmers in the Connecticut River valley that separates New Hampshire from Vermont and cuts through western Massachusetts.

Another way to look at our land is from the seat of a tractor as crops are grown and harvested. The first tractor I ever drove was at my grandparent Petersons' farm in nearby Fremont. It was a low-down red Farmall, not a tall green John Deere (which is the only kind Harveys drive). Seated on the bare steel seat, my feet couldn't reach the floorboard let alone the pedals. Grampy lifted me up there and engaged the manure spreader gears. He told me to drive around the field, following where the last load was spread, and he would "catch" me when I came back.

He and Uncle Philip filled another spreader while I was gone and indeed caught me by climbing onto the slow-moving tractor and steering it back to the manure pile.

They let me work with them all afternoon and I felt so grown up by the end of the day. That's another way to appreciate the land — through a mechanical interface. Though farmers are often judged by their equipment, it is a measure wasted on my family. Harveys historically eschew new machinery. We pride ourselves on making do with what we have, never buying anything new, always used, usually requiring a major merger of iron and hoses from multiple sources.

Early on, I appreciated the benefit of gas- and diesel-fueled equipment rolling over the ground, moving rocks, plowing, harrowing, mowing and conditioning hay, spreading manure.

Until my brother Matthew learned enough mechanics, Dad did all the baling because our rattletrap New Holland balers were forever shearing pins and stuffing themselves up. The knotters were unreliable and sometimes the mighty plunger hiccupped and rammed itself into the compaction chamber at precisely the wrong time and snapped the twine needles. Or something went wrong with the motor.

Dad would rush back to the shop and arc-weld broken parts back together. He spent many hours reaching into and lying under the old baler, often as dark clouds threatened to drench the exposed cow feed. We'd be loading trucks like crazy to get ahead of the storm, knowing full well that the race was already lost because the baler was bust. We'd catch up to where Dad worked and ask if there was any hope.

Inevitably, the sky dropped its load and soaked us, our clothing plastered to our skin, rivulets of salty hair-sweat water washing our faces and trickling down our spines. At that point we didn't care. We had done what we could and the job, such as it was, had ended for the day.

We kids weren't so numb as to never wonder why we never bought new equipment ... which leads to another way to consider our Epping land ... looking at it within the larger economy of United States agriculture. Friends in Iowa laugh at our attempts to cultivate such difficult soil.

Our machinery is ancient, compared with our western counterparts' huge modern tractors and crop equipment. But Mom explained it to us. She said we can't afford new machinery because it is manufactured with high-paid union labor and all we ever get for milk is $10 per hundred weight.

So we just have to "make do." Then, as a measure of sour grapes, she added: "Besides, those new tractors are so big they could never work on hills and around the edges of our irregular fields. They are built for square fields to go on the flat for half a mile in a straight line." So, who needs the new stuff, anyway?! We were born dependent on fuddy-duddy equipment because we couldn't afford anything better.

There's more to Applehurst Farm than minding our own business and simply working the land. There have always been the neighbors and trying to keep out of trouble with land-use regulators. (Our father didn't try to keep out of trouble. But the rest of us did (and usually still do).)

The sun rises over 500 wetland acres that Dad bought and resold with a tricky junkyard owner from Exeter. We lost friends during that episode involving speculation on hazardous waste and sludge disposal. Also to the east, just on the other side of Red Oak Hill, we sold about 70 acres for a development that turned out okay with attractive houses on big lots.

No such harmony with a Harvey homestead, northeast, where the new owner, Sheila LaBarre, murdered two men.

Then there is the Nottingham land to the north where Dad lost our shirts through litigation over minimum lot sizes. Plus the sale to a drug dealer who paid in small bills. And the spot where rustlers slaughtered our heifer. Northwest there is a hill that Dad proposed crushing for highway construction. Abutters were up in arms over that one.

And closer in, on the south, there was a neighbor named Marjorie Armstrong Coulter. I was fascinated by her dark blue eye shadow, bright-red lips and jowl patted with orange make up. My father taunted her and her meek husband, Lloyd. He insisted on circling their apple tree to head back for the next row of corn – tractor, chopper and dump truck. Inside the stalled truck, I cowered and cringed as Marjorie

Armstrong Coulter marched through her tall grass accosting Dad with what he later identified as a martini shaker.

She shouted and shook: “Mr. Harvey! Will you PLEASE refrain from driving on our property!!”

And what did my father respond? “Bless you, Mrs. Coulter! ... Bless you, Mrs. Coulter! ... Bless you, Mrs. Coulter!” He repeated this blessing for protection as he fiddled under the hood to restart the engine and retreat.

We never owned land to the south, toward town, where roads were already paved and houses occupied. We exercised no stewardship there. It wasn’t ours to care for. Still, we spent many hours overlooking municipal progress from the crest of our hill.

Hard landing, October 2008

Hangar with old orchard Fordson looking east.

The way we are ...

2. AND IT'S ABOUT A LIFESTYLE

Late Industrial Revolution technology launched progress — trucks and tractors, mechanical hay rakes, planters, plows and harrows, milking machines, refrigeration — the U.S. Department of Agriculture and Epping Planning Board — citizens surviving through interdependence: PROGRESS — good, bad, and ugly.

Entering the 20th century, it became obvious that not everyone had to be a farmer. This realization was a relief for those who had better things to do. Not everyone enjoys a good backache and the bracing whiff of manure. Most of the population realized there were easier ways to get a gallon of milk than to own land and cows that had to be dealt with every day of the year. No weekends off.

But we Harveys weren't that sharp. Neither were the Petersons or the McPhees or Sulloses or Allens. We Epping farmers just kept at it, growing crops to feed the cows so they'd make milk to sell so we could afford to plant again come spring to grow crops to feed to the cows so they'd make milk so ... "A way of life," we called it. Something we could do. Dad enjoyed being self employed. He called it "earning a baby's shirt."

Daniel Webster Harvey

Our friend Dudley Laufman explains what it takes…

TO KEEP A COW

If you put your mind to it
you could keep a cow or two
Have a parcel of land a pond
Pour a slab build a lean-to
off the side of the house
big enough for two Jersey tie-ups
storage space for fodder
five ton bedding too old hay
Clear some brush for a paddock
build a wooden fence
Cut green feed with a scythe
toss over the fence in summer
Manure to neighbor exchange for hay
Make lightly salted sour cream butter
cottage cheese to sell at the door
One Jersey might make 10 quarts a day
that's 2 and 1/2 gal which will yield about
a quart of cream = 1/2 # butter x 2 (the other cow)
that's a pound of butter a day
Skim milk to large curd cottage cheese
drink buttermilk whey on compost
Qt of milk weighs 2# x 10 qts = 20#
x 2 (other cow) = 40#
Feed 1 # grain per 10 # milk = 4 # grain per day
25 days to go through a bag @ $14
Say a bag a month app 50c a day
Get 5 bucks for the butter and cheese
that's $5 a day - the grain = $4.50
Don't figure the labor it's a lifestyle
supposed to be enriching
Like they say put your mind to it.

Reprinted, with permission, from WALKING STICKS (Beech River Books). In 2009, Dudley Laufman was awarded the nation's highest honor in the folk and traditional arts, the National Heritage Fellowship. Besides being a long-time friend of Petersons and Harveys, he is a musician and barn dance caller from Canterbury. A fancy plaque and $25,000 honorarium were bestowed by the

National Endowment for the Arts, celebrated with ceremonies and performances in Washington D.C.

Dudley works with his talented and personable partner Jacqueline. As "Two Fiddles," They conduct barn dances and residences in schools and community centers throughout New England.

"Get big or get out" was what we heard at 4-H and Cooperative Extension meetings and Farm Bureau and D.H.I.A. suppers. (That's "Dairy Herd Improvement Association," a select group of artificial inseminators.) The milk company also wanted us to "get big or get out." At first, we sold to a darling little company called "Clover Leaf Dairy" from Haverhill, Massachusetts.

"Johnny The Milkman" hauled our icy dripping-wet jugs for bottling and door-to-door delivery throughout North Shore, Massachusetts into Seacoast, New Hampshire. He dropped off wooden cases of half-pints at the school, glass bottles painted with green clover leafs. If Johnny saw me, he'd wave and make me blush. Everyone knew the school milk might have come from our farm.

Pretty soon, grocery stores began selling milk and home-delivery fell from fashion, forcing Clover Leaf Dairy out of business. Eventually, we ended up as suppliers to Agri-Mark, a company that charged $100/month to pick up the milk, backing in with a shiny tanker that hardly fit the driveway. Instead of former competition between Boston distributors and suppliers from New York State, Agri-Mark combined the two as a cooperative for all of New England and New York.

We were with Agri-Mark since 1980. Their stop fee was nearly offset by occasional big dents delivered by the hoary mulberry tree beside our mailbox. Ironically, Mom's concern for the tanker finally got my brother Matthew to cut down the mulberry last spring (2008) to avoid further damage. Six months later, there was no need for the milk truck to stop because Phyllis and her cows were gone.

I remember, as a child, Mom's lament that the price we received for a hundred pounds of milk was the same as it had been during the previous decade: $10/cwt through the 1950s and on into the '60s. Even today, nearly 35 years later, some might say it's even worse. They're talking about a per-hundred-weight drop from $12.50 to $8. Dairy farmers are still going out of business at an alarming rate with an occasional suicide thrown in for effect.

A Vermont senator is complaining that farmers receive less than a dollar for a gallon of milk, while consumers are paying three for it in stores.

Headline in Bangor Daily News: "Maine Milk Market Crashing, Farms' Income Could Drop 50%, Last two years prices lowest in 25," (1/14/09) An Agri-Mark rep named Doug DiMento (whatta handle!) is quoted: "This will put a lot of struggling farmers right over the edge. $11 milk is the end, the end of milk in the Northeast.'' He said lobbyists in Washington are pushing for another herd buy-out like in 1987. "We need to get rid of 250,000 to 300,000 cows nationwide to level supply and demand" to soften the "incredible cash crunch" expected next spring.

They say 2004 is the only year New Hampshire dairy farmers came closest to breaking even. The ONLY year.

Good thing Phyllis got rid of the cows when she did!

Too bad she had to be evicted by her family.

The way we are …

3. PHYLLIS ON NIGHT SHIFT

The Sisters' Refrain (please) ...

Phyllis and Beth sure have heart
But they don't seem to work very smart.
Our patience gets smaller
When in debt we waller.
This issue has torn us apart.

* * *

The customers stand aghast
'Til the threat of Arnold is past.
They can't weigh a potato
Or squeeze a tomato
Without being nosed in the… stomach.

My sister Phyllis has started milking cows at night, 20-minutes away from our farm in Epping. She and her 16 milkers were recently evicted by The Sisters — three of our sisters: Susie, Erica and Cindy — backed up by our parents who kept changing the subject.

"I told them," Phyllis said. "Got right in Daddy's face and said: 'Daddy. Listen to me! This isn't right what they're doing. What harm is it if I milk the cows at the farm? Isn't that what a farm's supposed to do … FARM?! I'm not making any money. But neither did you!"

She said, he just changed the subject, "As usual."

Now Phyllis leaves her house after dark for the 9 p.m. to 2 a.m. shift at a dairy farm in Lee. Another sister, Beth, goes with her.

That's all of us six sisters: Susie, Phyllis, Cindy, Beth, Erica, in that order with me, the oldest before Susie. We also have two muscular blue-eyed brothers: Matthew and Albert. They are the only ones who came to our son's wedding last June, them, their kids, and our parents. One hundred in attendance and not one of my sisters could be bothered. Can't say I was happy about that. Phyllis tried to come, but couldn't leave her dog.

Phyllis and Beth share a wage of $10 per hour with $2 of it withheld by the employer for some sort of "bonus." I don't understand how that works. We Harveys have never been known for negotiating heroic employee benefit packages. But still!

Phyllis's dog goes, too, and runs free among the herd of 150 Holsteins. I don't know why anyone would want a dog like 'Arnold'! The creature has no apparent manners. Once, he came into my house and went directly to the fireplace to take a piss. Striking black-and-white with stocky pit-bull conformation, Arnold has alleged herding talent.

Over time, Arnold has become more tolerable – mellowed by age, like the rest of us. But before being released from the inch-thick rope of his youth, he was vicious and bit the cows' noses if they came too close. He went berserk if he got a shot at a cow's flank, sure sign of herding instinct — a flanker, not a heeler.

Outside the barn, he lunged at and clawed people while choking at the end of his tether. Phyllis would say: "Oh, he's all right. There! There! Arnold. Settle down." She let him jump on people's cars, which was seldom appreciated. She left his rope long so he could sleep on top of our father's fifth-hand Mercury Cougar.

Vegetable customers stopped at the stand, sometimes riding motorcycles. It drove Arnold nuts. He HATES motorcycles and ran so fast toward them that the rope would snap taut and flip him onto his back. Given a chance, he bit into their tires.

Our neighbor Kim arrived and specifically asked Phyllis: "Is Arnold going to bother me when I start this motorcycle?"

"Oh, no! He's good. Aren't you Arnold?" my sister said.

Kim started the machine and Arnold chomped into his calf.

What's not to like about a dog like that?

Anyway, now off his rope, Arnold is on the night shift with Phyllis and Beth, roaming among cows that someone thinks are worth milking three times a day (instead of the standard twice). Hopefully, the dog won't chomp their calves.

Part One - Terminated

My biggest gripe with Arnold is that he is with Phyllis ALL the time. Now that he is off the rope, she won't leave him tied or penned at her house because... Well, I don't know why. I guess Arnold freaks out or Phyllis freaks out — one or the other or both — I don't know! But restraining him is no longer an option. There is no kennel and I doubt she even saved the rope that once bound them together.

A mere month ago, my weather-beaten sister was talking sweetly and milking her 16 cows inside our family's six-on-a-side herringbone-design parlor, as she had for the past 12 years.

And now, she's splitting $10 an hour, no, $8 to work nights at a farm owned by people we don't know. And it takes five hours to get the milk out of them! FIVE hours to milk 150 cows! Even Phyllis says it is "very slow." I tell her it's crazy, since I milk cows in New Zealand where it takes five hours to milk 400 cows TWICE a day — both milkings (morning and afternoon). Five hours. Over. Done. "Home and hosed!" A hundred bucks in my hand, Thanks!

In 1987 when the Harvey herd was first slaughtered, I said: "Good! I'll never have to milk a cow again!"

Yet, I recently resumed milking in New Zealand more as a joke than anything else. I had noticed "Assistant Herd Manager" listed as a work permit possibility – a dirty job that New Zealanders leave for foreigners. Not only was it something I was qualified to do, but just maybe it would score me points back in Epping.

You see, I lost credibility as a Harvey 25 years ago when I left the farm. So, I figured if I could milk more New Zealand cows in a day than were ever milked in a fortnight back in Epping, then maybe my family would notice.

But it didn't work. Each year, (eight years now), I return to the Epping farm after traveling around the world, absent for six months, and they don't even ask where I've been, let along marvel at my milking experience.

Good joke on me! I roll out of bed every day at 4 a.m. to become totally splattered with manure and my family just thinks I'm stupid for dealing with cows.

Not a problem, really. Me milking cows unnoticed or Phyllis doing it for peanuts. Not big enough problems to appear on the current list of problems at Applehurst Farm. As long as Phyllis stays over on her side of the fence and I go away ... and Beth, too ... then we're not in The Sisters' way and our parents can change the subject all they want to.

Phyllis is "just sick" about being evicted. But she will have to heal like the rest of us. She says she understands that The Sisters wanted more rent than she could afford to pay. She claims to be grateful "befriending" another herd and milking in a modern facility.

"Just sick. Yes! I'm sick!" she laughs. "I love cows and a job's a job. I just wish things were different and that Applehurst was a proud dairy. I'm not convinced it's impossible. But, today, farming is looking pretty grim."

Matthew digging Alfred Carr's grave in family cemetery.
Tombstone epithet: "A Good Cow Man" (per Mom's instruction).

Arnold on Dad's car

The way we were ...

4. LATER, SWEETIE PIE!

My father often kissed Mom good-bye after announcing where he was headed: downtown or to see his friends Arnold and Otis or to a meeting. He'd give her a little hug and smooch, saying, "See you later, Sweetie Pie!" Or, "I'll be back, Honey Bunch!" Sixty years later, he still tries to do the right thing, though it hasn't been easy living up to Mom's expectations for punctuality, production and generally not goofing around.

Mom shooed me off a hundred times: "Go tell Daddy it's getting late. We've gotta go. Pull his sleeve."

We'd all be crowded and fidgeting inside the sweltering (or freezing) station wagon outside any possibility of buildings where Dad just couldn't help himself. I could kick him in the leg and he wouldn't flinch. He just kept talking and talking and talking, usually about politics and why he's 'against' the Epping and Nottingham Planning Boards.

That's what he'd say: "I've got four (five, six, seven or eight) kids. I'm raising them to be Democrats and against the Planning Board."

He always wanted to be on 'the other side' whatever the other side happened to be. Since all of my relatives — on both my parents' sides of the family — seem to be Republicans, I've often wondered if Dad declared himself to be a Democrat just to be on 'the other side'. I think he'd make a pretty good Republican, a conservative by instinct. True to Cold War mentality, Dad said the planners "got their directives indirectly from Moscow," which was supposed to mean our government was turning Communist.

Forever trying to outfox the newcomer land-use controllers, my father preached civil disobedience. It was a pitched battle, us against the planners. Dad told everyone to ignore laws while he himself was a stickler for legal protocol, especially the prescribed posting of notices and advertising for zoning ordinances. Dad complained if some letter of the law was not followed. He became strident and defensive, decrying the "police state" while demanding his right to petition as noted in 'The Constitution' and 'The Bill of

Rights'. He called the planning board the "prevention board" and "monkeys chattering in the trees" and "dogs in the manger" who wouldn't let the cows eat their hay. He said the voting public were "sheep" seeking protection from "dangers that they know naught."

"Mention unconstitutionality and people go psychotic and try to run up a tree," he crowed.

It's an odd thing about Dad. He will give you the proverbial 'shirt off his back' if you request it. But don't ask him to give up local politics or try to tell him how to manage his land because he will do the exact opposite.

Growing up in a house with an opinionated Republican type claiming to be a Democrat was bound to warp us kids. We developed a critical attitude toward government and believed we were poor — on the verge of having the Town take the farm for back taxes or maybe a bank would call in our loans. It was an unfortunate way to think, since we always had plenty to eat and we wore good enough clothes. Our blessings were many.

We were NOT poor! We just thought we were.

Dad always said we were 'honorable' because we are farmers and know how to take care of land. He said if it weren't for us, people would go hungry. He said our cows were the "mothers of mankind." Still, it felt like wolves were at our door — bankers, feed and fertilizer salesmen, code enforcers, tax collectors — and nobody, not even our friends, felt sorry for us.

You see, we Harveys have always believed we deserve more respect than anyone else would ever guess they owe us. Dad claimed to fight for the rights of the 'common man'. He felt that we, along with his imaginary 'common man' friend, were underdogs oppressed by the economy and demands of people who had recently moved up from Massachusetts.

After the Epping planning board was reauthorized by the selectmen in 1967, Dad strengthened his bonds with the natives, the downtown men and women he grew up with. Since he is likable enough, these old friends gladly signed Dad's petitions.

Subscribing scores of two-acre houselots didn't add up to much on Dad's tally and must've made his friends wonder what all the fuss was about. If Dad collected signatures from owners of 20 percent of Epping's land area, then land-use regulations had to be adopted by a two-thirds majority vote, rather than just a simple majority.

In most cases, Epping's 'common man' had more in common with newcomers from Massachusetts than with the talkative farmer on the hill. Truth is NOBODY in town, except the three other farmers, owned nearly as much land as we did. The landowner petition would have succeeded with just the farmers' signatures. Still, Dad's petition was popular and many people wanted to be included in the 'revolt'.

My father's grassroots tactic effectively delayed Epping's 'master plan' and zoning for ten years, totally frustrating local and regional officials who wanted to get on with PROGRESS.

In 1978, the planners got smart and did an end-run around Dad's petition. New Hampshire legislators ruled that warrant articles proposing to abolish planning boards were amendments to proposed zoning ordinances; thereby also subject to a two-thirds majority vote and, in the end, made zoning untouchable.

I dare say Dad's court cases and political jujitsu worked against the farm during its most difficult decades, mid-'60s to mid-'80s, when there wasn't enough money to go around and animosity intensified at Town Hall.

So what did I do right in the middle of this? As a recent UNH graduate, I went to the Epping Planning Board and applied for a building permit. Such a disappointment to Dad! He had paid good money to put his oldest daughter through college and here she was going 'hat in hand' begging for permission from The Enemy. Hadn't I learned ANYTHING! Maybe I had caught UNH Communism, just as Uncle Eddie said I would.

Besides dealing with local regulators, I made my way through bureaucracy in Concord where I was issued the state's first compost toilet permit.

It wasn't much fun dealing with the government while taking guff from Dad. Still, I was determined not to live in fear of inevitably being caught, even if it meant I was a 'sheep'.

The planners made me cry during a meeting. They said I needed frontage along Nottingham Square Road and asked why didn't my father deed me a piece of land? I told them he refused to cooperate.

"Why?" they wanted to know.

"I guess he doesn't like me enough," I sobbed and they understood because he didn't like them either.

Taking pity on me, the Board caved and sanctioned my building permit while Dad eventually deeded the 30-acre frontage to my sister Susie. He put her name on land between me and the road, bounded just 50 feet from our house, for the same reason he withheld it from me. He didn't want to deal with the planning board and refused to subdivide, first for my access, then for my sister's roadside rental house lot.

I remain "land locked," a consequence of buying from our friends The McPhees, rather than choosing a piece of Applehurst Farm. Still, by following town ordinances, I haven't had to worry about being illegal since 1971.

Dad never stopped fighting land use restrictions.

When I think back on those years, a particular gem of correspondence stands out. A simple message typed on Epping town letterhead: "Why, Dan Harvey? Why?!" is all it said, signed by Peter Planner (not his real name) after yet another failed zoning referendum.

The Planning Board chairman was not well liked by Dad's team, so the letter was seen as a trophy. Uncle Eddie (Remember, he's Mom's sister Ann's husband and Dad's sometimes ally) laughed when he read it.

"That guy's so slippery he has trouble keeping his socks up," Ed said. They scoffed Mr. Planner's profession: Public Relations, saying it was nothing more than being a "lobbyist" or a "con man." The words were spoken with disdain.

Dad said Peter Planner was good at making black look like white. "He makes speeches about how the planning board has no power, but all along he's scheming to control everything that moves." One year, in response, Planning Board members wore "NO POWER" t-shirts to Town Meeting.

We Harvey kids were young and impressionable while these antics were going on. As we entered public school, it was hard to understand our place within the community. With so many opinions floating around, it was confusing. We had been brought up on the farm, which was odd enough, since even the other local farmers weren't impressed with our ability. Plus, classmates didn't understand why we often smelled like wood smoke, animals and silage.

Part One - Terminated

Our parents seemed to provide two different distinct role models. Mom obeyed society's rules; Dad did not. So, since we love and admire them both, it was unclear who to follow: Mom or Dad.

Married February 4, 1947

The way we were ...

5. CARPENTRY BY CHARLIE STROUT

Our house burned flat on my grandfather's fourth birthday, in 1874. Therefore, the actual house is over 60 years newer than its foundation. The three-story structure is nothing fancy, a straightforward farm house, white with green trim. It has side-door porch entry and a little roof over the front door.

The ground floor consists of a big kitchen, dining room and a hallway wide enough to accommodate viewing Grandmother Harvey's coffin in 1965. Mom papered the walls red for the occasion. Very pretty for gladiola bouquets.

A flight of stairs accesses the second floor from the hall. There are 15 steps. The second one from the top squeaks and you need to avoid it while sneaking in late at night. Climb more stairs to the attic where there is plenty of headroom because the kneewalls are four feet tall. Mom made two more bedrooms up there for Matthew and Albert as it became crowded on the lower levels.

Sometime before 1776, the original house was built by the Ladd Family. According to records, my great-great grandfather James Harvey moved from nearby Pawtuckaway Mountain in Nottingham to marry Lois Ladd in 1813.

Our ancestors hit ledge while digging the foundation. Then and now, Harveys believe in full cellars (meaning dug deep enough to provide storage space without having to crawl around to use it). So, with bedrock always a problem, the old-timers laid up an above-ground stone foundation and filled around it. As a result, the house and its extension sit on a rise about six feet higher than the road out front — though the pavement gets higher with each application of tar.

My parents were dead set against having the road paved in the 1950s. Mom said it needed to remain dirt so the cows and we kids wouldn't be hit by cars. The parents went to Nova Scotia one week, and SURPRISE! Suddenly the town road crew showed up with a grader and tar truck. As the oldest, I frantically wondered whether to drive the men away with a pitch fork or serve them Kool-Aid. There was no way to phone Mom and Dad.

After the 1874 fire, my great-grandparents had the "foresight" to rebuild the chimney with staggered strips of lead flashing hanging out 12 feet above the roof of a section we called "Grammy's Side." The thought was that "someday, someone" would add a second story and the flashing would already be in place. This irregular collar of gray metal halfway up the chimney served as a talking point until I was well into high school.

Finally, in 1962 (around the time I got my driver's license), Mom decided it was time to take advantage of Great Granddad's foresight. She sketched a major expansion to the house. The plan was to add three bedrooms over 'Grammy's Side', since my grandmother had left to live with Aunt Edna and Uncle Bill Woodward in Durham.

Dad never claimed to be a competent carpenter, though he does pretty well lining up locust posts for pole sheds. So he was off Mom's hook. She boldly hired Charlie Strout, a professional — though ancient — carpenter from Elm Street, Epping.

Mom, Susie and I catered to the various needs of this elderly French-Canadian builder who was quite frail, deaf, and nearly blind. He stuttered while talking loudly to himself. And he couldn't drive a car, so I picked him up in the morning and drove him home at night. Plus, he couldn't lift anything very heavy, so either Mom or I remained available to hoist. If he dropped a nail or something small, he couldn't see it. So we stood by for that, too. But he was a good carpenter who just tapped his head whenever someone asked to see the plan.

"It's all in here!" he said. The project seemed to last for years, but it was probably just a summer, since I don't recall being snowed on. I learned the basics of carpentry while helping Mr. Strout.

Surely, our forebears would have been pleased with our work and Mom's determination that there should be an upstairs toilet and shower and we should each have our own bedroom.

Susie, Phyllis and Cindy moved into the new wing and were encouraged to decorate any way they wanted. Phyllis's scheme was the best. Space exploration was big then. She painted her walls as a moonscape with waist-high craters and bright blue sky overhead. The Big Dipper and other glowing phosphorescent stars were quite convincing. One wall was black featuring Earth in the distance as a blue-green orb. Later, when Beth, (the older of the two 'Babies') took over the Space Room, Phyllis helped her add Neil Armstrong in his space suit holding an American flag.

The other two rooms — Susie's and Erica's — were unremarkable with tiny-floral wallpaper.

Aunt Edna assembled a scrapbook of old photos and papers indicating that my grandfather and great-grandfather were good farmers and involved with local politics. So, Dad comes by it honestly. Great-Grandfather Matthew James Harvey wrote poetry and served two terms as an elected state representative, while my grandfather (Matthew - no middle name) was known to be "extremely tolerant" and helpful to the needy.

Mom says her father-in-law's tolerance encouraged "undesirables" (probably drunks) to hang around the farm, trading labor for hard cider. I suspect she shooed these men away as soon as Dad inherited the farm. My grandfather was Epping's moderator for town meetings and a tax collector, plus overseer of the poor. I never knew him because his heart failed when I was just a couple of months old.

My ancestors kept a few dairy cows, but were primarily apple growers. They exported Baldwins to London, packed inside large wooden barrels. As I recall (and I could be wrong), my Grammy (Alice) Harvey also sent things to England. She raised chinchillas for fur and shipped their slinky pelts overseas. I remember her battery of cages and the weasely little noses poking out of them. Unfortunately, that's all I recall, except that she read pilot books, continually knitted itchy mittens, and ran her upright barrel stove so hot that the sides glowed red.

Covered cowyard under construction. Dad lined up locust posts. Matthew 'custom-installed' the roof, meaning it's not easy to install a straight ridge line on crooked posts.

House we grew up in. Middle section above 'Grammy's Side' was raised for three second-story bedrooms, built by Mom and Charlie Strout. Matthew and Albert slept in the attic.

Tank truck collecting milk (probably driven by Dick Bard). Left to right buildings: Chicken house, milking parlor, milk room, horse barn. Front: Well house. Mom's vegetable stand is buried under snow by tree to left.

Uncle John and Aunt Agnes Gilbert (neither a blood relative) at Cindy's barn. David & Cindy (l&r).

Me pointing north from top of Red Oak Hill over our last crop of silage corn. After Whole Herd Termination (in 1986-87) Susie and her husband Tom grew sweet corn for sale throughout Rockingham County, requiring much smaller acreage than what we used to plant for the cows.

Part One - Terminated

The way we were ...

6. MOM. A NANNY?

Dad's enthusiasm for cows diminished during the 1960s. Our herd had grown to 100 milkers with about as many young stock. It was as big as we would grow, the peak of our milk production, the top of our game as a family team.

We had installed a New Zealand-style milking parlor (herringbone shed, six on a side) and expanded the dairy operation past the point of McPhees and Allens ever catching up, while the only other Epping dairyman, Jim Sullos, continued milking 15 cows by hand.

Dad said he had "pushed the business" as far as it would go. There, at the limit of our credit line, we couldn't afford to hire anyone to do the milking. So we took advantage of any volunteers who came along and did nearly all of the work ourselves. Matthew did most of the heavy lifting, planting, harvesting and feeding-out the forage crops. He also upgraded the facilities with expanded roofs, cow yards and trench silos. His five-years-less-experienced brother Albert helped maintain the trucks, tractors, choppers, manure spreaders, planters and other equipment. It was early training for his later career as a mechanic.

We took turns milking.

Though it felt like a fast-moving assembly line at the time, the small scale of just six sets of cups was quite manageable. So much easier than the old tie-up. No more backache from bending to milk, constantly up and down, plus lugging jugs of milk to the milk room. In theory, the cows could "run through" the parlor at a rate of one cow per minute — six minutes to milk a side, six cows on a side — 100 cows in less than two hours. (Using the same arithmetic, it takes me 15-20 seconds to milk a cow in New Zealand — 400 in two hours.)

The "beauty" of the milking parlor — as Dad crowed about it to the McPhees and Allens — the "beauty" was that cows walked in on raised platforms so their udders arrived at chest level. No more bending over. And, wasn't it "beautiful" the way individual servings of grain dropped onto the cows' heads when they stuck their noses into the hoppers! The fancy glass pipeline promptly cooled, strained and dumped milk into the beautiful bulk tank. Our cows were happy and the workers might have been thrilled.

Oh, it was quite an improvement, all right! We should've been clambering over each other to try our hands at milking. Dad declared that "just one person" could easily do the lot. And, if he could manage it, that one person would NOT be him.

Lo these many years later, I can appreciate Dad's lack of enthusiasm. Having completed hundreds of milkings myself, I cannot recall showing up even once with feelings of "O Joy! O Bliss! Aren't I lucky to be milking these cows!!"

The only person who seems to enjoy the process is the herd owner – here in New Zealand, not back in Epping – and that's because he sells $4,000-worth of product A DAY … which prompts an observation on "economy of scale": If a cow provides $10-worth of milk nine months of the year, it makes no economic sense whatsoever to milk a handful of cows (like Phyllis's two or even 20). Two cows = $20/day for nine months. Not enough to sustain a dairy.

I can't remember the figures or frequency of payment back home, but Applehurst's milk check lacked zeros on the end and (in the best of times) grossed around $100,000 for the year.

Here, at least I earn a decent hourly wage. Back in Epping, even my parents made bunkus for profit.

I was a sophomore in high school when Mom threatened to leave home. As the oldest of six, fretting with adolescent hormones, I was terrified by Mom's suggestion that 'we could all take care of ourselves'.

"No. We can NOT!" I declared as I followed her out to the barn to milk the miserable cows. "I can't even take care of myself!" I muttered while slamming gates and yanking the six sets of teat cups and black rubber hoses from where they soaked in the stainless tub. Mom had stormed off toward the pole barn to collect the herd while I set up the shed.

Dad really blew it this time! He had passed Mom on the road at milking time. He was driving toward town, away from the farm, while Mom was headed home. It really upset her that he waved at her.

Before our march to the barn, Mom had sobbed: "That's it! ... He just WAVED!" She moaned that she never meant to be stuck milking the cows. "HE's supposed to do it.

Look at what TIME it is! And he just waves!" She told me she couldn't stand it anymore and that we kids were big enough now to take care of each other ... and she was leaving, which started me crying, too. I asked where she would go.

She said: "Boston."

"What will you do there?" I asked and made her even more upset.

"What CAN I do?" she sniveled. "I'll work as a nanny. I can take care of children. That's all I know how to do." She pulled a smelly sweatshirt from the barn clothes rack, slipped it over her glasses and threw thc door open to go deal with the cows.

Tears blurred my vision as I finished assembling the milking machines. I envisioned Dad without Mom. He'd be SORRY! And all us kids without a mother. I was missing Mom already and feeling acutely sorry for myself when, suddenly, Dad appeared at the door.

My damp bloated face did little to dispel his worst fear: that he had really screwed up and it was a good thing he turned around and came back.

"Mommy's mad," I said and brushed past him without further explanation. It felt like I would explode with emotions that I did not want to reveal to Dad. My stomach ached and my brain buzzed with the decision to exit that scene as fast as possible. I began running out through the barns — the 'New' one, then the 'Old' one, into the trench silo where I scaled the back wall to slip through the repair shop and escape to the junk.

My heart pounded — not so much from running as from pain in my head. My brain pulsed with anxiety. I needed to get out of sight ... quickly.

Quick! Through the shop and then I'd hide somewhere in the junkyard. I'd ... almost fall over UNCLE EDDIE who was repairing a tractor tire in the shop.

"AAAAhhhhhh!!!" I shrieked as he rose from the shadows.

"Slow down, Girl!!" he said in something softer than his usual matter-of-fact voice.

"I-I-I've gotta go!" I blubbered, unable to staunch my breaching flood of tears.

He sat me on a grease bucket and asked what was wrong. I couldn't talk because I couldn't breathe, having reverted to my habit of spasmodic shallow gasps. My uncle waited patiently.

"OK, now. Tell me what happened," he said and sounded ready to listen.

So I told him about Mom saying she couldn't take it anymore and was leaving to be a nanny, and that I had left her and Dad to do the milking.

"I'm scared," I said.

"What are you scared of?" my ever-present uncle asked.

I told him I was afraid for Mom, that she'd become lost in Boston.

But, really, I was scared for myself. I didn't want to be responsible. I would be lost without my mother and I could already feel the farm slipping away.

Picture taken in Applehurst orchard sometime before my grandfather died. Handsome bloke on right is Dad supervising well-dressed pickers from Watson Academy, no doubt.

Part One - Terminated

The way we were ...

7. NO ROASTED PIGEONS FOR MOM

Besides an airplane and a new potato planter, back around 1946, my grandfather Matthew Harvey gave Dad another agricultural enticement.

The real clincher!

He said: "Go see about that Peterson girl. She's the kind of woman who doesn't expect pigeons to fly down her throat already roasted." Dad was 26; Mom 21.

A year later, my parents had married, I was born, and my grandfather died from flu and heart failure at age 78.

While Dad's obsession was local politics, Mom couldn't be bothered with Town Hall. Her genuine interest centered on other aspects of life, primarily around raising our family. She made us proud, never embarrassing us in front of our friends, always putting good food on the table: homemade whole-grain and Anadama (cornmeal & molasses) bread, baked beans, chicken pot pies and casseroles, soup with dumplings; desserts galore: apple, blueberry and lemon meringue pies.

Though she can cook most anything, I have to say 'boiled dinner' (made with good ham) and fish chowder are her signature dishes. We knew nothing of pizza. Our son Otis was weaned on Grammy's yeast rolls.

She is an expert seamstress, good at patchwork and tailoring. She used to make all of our clothes – most memorably, wild-pattern corduroy pants. Mom knows how to cane chairs and do upholstery, and, of course, competent with painting and wallpapering. For awhile, she made furniture. A piece featured in our dining room was always called 'The Thing Mommy Made'. It was a bureau with shelves on top showcasing indispensable treasures like a painting Great Aunt Muriel did of Brookvale Farm and a ceramic plate I created depicting three of my calves: 'Pearl', '10-1/2', and (the freemartin) 'Pepper'.

Our mother has a large loom and gives away scads of rag rugs. Springtime and summer, Mom lives for her flower and vegetable gardens. Even now, at age 85, she rototills in her bare feet and happily bestows her tomatoes, zucchinis and zinnias upon visitors from downtown and beyond.

When we six older offspring were pretty well grown, 'The Babies' Beth and Erica were born. Undaunted, Mom decided she could work around the kids and start earning a paycheck. She drove a school bus for several years, as did her sisters, my Aunts Alberta and Ann. Then she got a job delivering Meals on Wheels, cheerfully keeping an eye on scores of housebound citizens. She was perfect for the task and saved at least one life by arriving during a crisis. Mom still maintains her connections with that service network. At some point, she went to work for Joan's Flower Shed in West Epping where she mostly delivered flowers throughout Rockingham County, but also did some arranging.

Asked about politics, Mom would say: "I have better things to do," and nobody could doubt it for a second. Neither could one miss her sentiment that Dad's preoccupation with Town Hall was a distraction from farming — though she seldom complained. She DID occasionally take offense when someone said mean things about Dad. Even then, it wasn't so much the politics that tweaked her as having comments made to her directly, as if she wouldn't always defend her husband. She turned a cold shoulder to the critics until she figured they had learned their lesson (which may not have been the case).

Mom discussed anything BUT local politics even as she became one of Epping's most active Parent Teachers Association (PTA) members and a deaconess for Congregationalists in the Community Church. She was a key player in serving church suppers and the annual chicken barbecue. She also volunteered as 'classroom mother' for each of us while we progressed through Epping Elementary School. She organized parties and chaperoned our field trips. She led 'Hilltoppers' 4-H Club.

Mom often sent us off on the bus with slices of apple pie for our teachers, packed in wedge-shaped Tupperware — the pie, not the teachers. I remember once she put too much cinnamon in Mr. Warren's pie and I was mortified because it was the same day my friend Sherry Chase told him I said he looked like my horse 'Ginger' (which, in my mind, was a compliment).

Despite politics, Mom was elected to the school board and was a voter checklist supervisor for many years. As a community volunteer, she convened the first meetings of what is now Epping's Conservation Commission. She coordinated Epping's 4-H clubs to pick up trash along the roads. We collected tons of beer bottles in early spring, sometimes plucked from ice under trees, usually before poison ivy and black flies emerged. But not always.

Through the PTA, she and other mothers established the town's first hot-lunch program, using facilities at the nearby Grange Hall, where she had pull as a lifelong member. At noon, the whole school walked over there to eat. The Grange had (and still has) a serving kitchen and assembly hall. For several years, Mom was responsible for preparing desserts for one hundred kids: blond and chocolate brownies, applesauce cake and squares, peanut butter & oatmeal cookies, jam-filled cupcakes, things like that — all baked in our farm kitchen, made mostly from government-surplus ingredients such as powdered milk and eggs, peanut butter, desiccated fruit, flour and all sorts of other not-readily-understood baking supplies. Even the bag-lunch kids got Mom's desserts, since they had to eat at the Grange anyway to be supervised.

Our mother has always been an efficient and resourceful homemaker, clever with decorating and maintenance, a good hostess. I've always appreciated that Mom never forgets my birthday (or any of the other sevens'). She used to organize field-trip parties for me, inviting my classmates, usually to climb Pawtuckaway Mountain in nearby Nottingham amid autumn foliage.

Mom raised us without physical punishment. Neither she nor Dad ever raised a hand to us.

I don't remember what I did to nearly get a spanking. But it was a matter of: "Daddy will deal with you when he gets back!" And when he announced that we were "going to the woodshed," I was embarrassed to tears as my father sat on the toilet, flopped me over his lap and slapped the bathroom wall. He told me to holler.

Mom grew up different from Dad. She and my Peterson aunts and cousins still work hard. I think they'd go adrift if they didn't.

They also do a fair job of holding grudges. Damn the day J.F. Brown and Sons, Epping's International Harvester dealer, sold my grandparents a defective refrigerator! Unaccustomed to servicing home appliances, the tractor mechanics kept trying to fix the compressor. Or maybe it was something as simple as the door hinge or gasket. I forget. But whatever the problem, the Browns lost a good customer — for life — and their name was never to be spoken at Brookvale Farm.

One good thing about my mother marrying Dad is the way her attitude is tempered by the Harveys' tendency to forget and not sweat the small stuff.

Everyone said Grampy Peterson was a good 'stock man', that he had 'a way with animals'. This was true. Cows, especially, loved my grandfather and responded to his voice and gestures. He was quick to scratch behind their ears and rub the sweet spot under their throats. Yet Grampy was not to be trifled with for more difficult animal behavior — kicking while being milked, breaking through fences, thrashing round while having their hoofs trimmed or during physical exams. Grampy knew how to growl at the cows and carried a walking stick for an occasional whack.

People also said he had a good sense of humor. But perhaps it was too refined for children, since I can't recall any of his jokes. Indeed, I don't remember him laughing much at all, most likely because he wasn't feeling well. I didn't even know he was sick until he died. I remember that he refused to take aspirin.

"Nope!" they'd say. "Albert won't take aspirin, no matter how much he hurts." It was as if suffering without relief was a virtue. Still, I recall being fed plenty of orange-flavored 'baby' aspirins. So what was the big deal?

As I saw it, my Nova Scotian-Swedish grandparents were loving and attentive to me as their first grandchild. But I also appreciated that Grammy had a short fuse. She came to live with us for a couple weeks when Mom nearly died after birthing one of us. I don't remember which baby, but Mom was so sick that she became incoherent and Doctor Blaisdell made a house call. I don't know what we were doing that was so bad, but Grammy even yelled at me, her favorite.

"If you kids don't behave, you may not have a mother much longer!" It was SUCH a 'Grammy' comment.

I know it's true that Mom nearly died because she described the experience later.

"It looked pretty good," she recalled. "I was standing in the middle of a dirt road with full-blossom apple trees on both sides. They were beautiful. There was a little bridge crossing an underpass like where the cows go through to the back Plumer Field." She said old Plum Ladd was waving to her from the field, encouraging her to come on over because the sun felt so good and the blossoms and wildflowers were sweet.

Plum, of course, had died several years before and was missed because he had visited frequently and always tooted the horn when he drove by. Mom said she decided not to cross over. She wanted to stay with us.

Grammy Peterson considered herself a solid judge of character. If you didn't get on her good side right off, you probably never would. She was especially critical of our various spouses. She wasn't above playing favorites to get the most work out of people, even — or perhaps, especially — idle bystanders. She said "bah!" and "hah!" often and commented on who was "just standing around doing nuthin'." It was the most offensive thing you could do: stand around.

Nevertheless, Grammy and I had many good times together. I cherish memories of her holding my barf bag during our first flight to Fort Worth and then defending me against the vagaries of her even-more-temperamental sister, Muriel, my great aunt who often referred to me as "that little brat." Indeed, Grammy and 'Auntie' influenced my decision to go to college in Texas as a first time away from home.

I enjoyed hearing Grammy laugh. When something "tickled" her, she trilled like a Red-winged Blackbird in springtime.

We bought a wig together. A RED one. She and I as partners in the $30 investment. Grammy wore it more than I did because it was her natural color, not mine. She told me to never wear mascara, while pointing out girls who did: "Just look at that mess!" (Damn their eyes!) She also told me not to sit on rocks or I'd get hemorrhoids.

Grammy Peterson taught me all sorts of good stuff, including how to work in her vegetable garden.

She took me out to eat a hundred times. Grammy just loved the 'Route 125 Restaurant' in Lee and always referred to it as 'The One Hundred and Twenty-Five'.

Later in life when I was old enough to join her in a cup of coffee, she complained that a cute little breakfast place I took her to in Exeter served coffee "bitter as gall." She didn't appreciate the decorative sprinkle of confectioner's sugar on her French toast and cast aside the fresh orange wedge garnish, exclaiming that the eatery wasn't anywhere as good as 'The One Hundred and Twenty-Five'.

Grammy and I ran errands and kept doctor appointments together. All through second grade, once a week, I went with my grandmother to an osteopath in Dover. At first, it was therapy after I broke my arm. But I soon got over it and mentioned to Mom that I wanted to stop having my neck snapped for no good reason. She said I had to keep going for Grammy's sake.

"It really helps her rheumatism and she won't go without you," Mom said. So we kept going even though I grew increasingly tense before each neck twist and the doctor had to try multiple times to get a good crack.

The things we do for relatives!

Sawmill collapsed one winter; airplane hanger on horizon; salvaged fuel tanks foreground.

Grammy Eunice Peterson helped with Baldwin cider apple crop. One small orchard block was spared destruction. Fortunate great-grandbaby is probably Paul and Cindy's 'Abbie'.

Grammy Peterson died August 14, 1996, buried in Sleeper Cemetery on Brookvale Farm. Grampy Albert died January 1, 1955. Uncle Philip's cremains are also there (buried in a milk jug) following death on September 28, 2008. Photo from Grammy's funeral: My Brother Albert with Cousins Eric Schreiber and Dana Burtt. Background: Peterson barn built in 1945 by our Grandfather Albert.

The way we are …

8. THE NIGHT PASTURE

The Sisters' Refrain (please) ...

> ***Our father always liked land***
> ***We thought farming went hand in hand.***
> ***But we got tired of losing***
> ***And had to start choosing***
> ***So drawn was a line in the sand.***

* * *

> ***Rise and spread the alarm***
> ***We think we're coming to harm!***
> ***Forget about how***
> ***She handles a cow.***
> ***Phyllis is milking the farm.***

Encircling the barns is a rolling 15-acre field we call the "Night Pasture," though it hasn't grazed cattle much at all, mostly just during the 1950s and early '60s.

Probably back in the old days before The American Revolution, it was forest — mixed hardwoods, birch, poplar, maple, red oak — same stuff that continues growing back today along the edges. Back then, as the 18th turned to the 19th Century, my ancestors cleared it and piled the glacial stones along what became town roads and a cow lane that directed stock across the Night Pasture to an underpass leading to more westerly grazing on the other side of Nottingham Square Road.

No doubt, back then, a few milkers and calves were fenced for home supply milk, butter, cheese and meat.

My great-grandfather, Matthew James Harvey, planted an extensive apple orchard in the 1800s. Then, my Grandfather Matthew managed the business while increasing the farm's acreage through purchases of clear-cut forest bought cheap from our sawmilling relatives, The Ladds. This practice of acquiring unattractive ledgy land continued when my father inherited the farm in 1947. He bought many parcels from our elderly neighbor 'Plum' Ladd until Harvey holdings exceeded 2,000 acres.

As soon as his father died, Dad began destroying the orchard in favor of cows. He believed apple trees and cows were incompatible everyone knew it ruined cows to eat apples. (They get drunk, bloated, and their milk dries up.)

At first, the herd was small enough to be turned loose to graze. The Holsteins and Jersey- and Ayrshire-crosses enjoyed the run of the farm. After morning milking, they foraged in fields named 'Across the Road' and 'Through the Underpass' and 'The Plumer Field' and 'Edgerlys'' and 'Crockers'' and 'Plum Ladd's' and 'The Orchard Field'. At night, they slept close-by in the 'Night Pasture'.

As herd size increased during the 1960s, it was decided to switch to a feedlot system, where the cows remained undercover year-round. My brother Matthew expanded the professional construction to feature a much larger roof over the feed alley and four rows of 'comfort stalls' where the cows could lounge on sawdust while not lined up to eat hay, grain and silage from the long manger. Each cow had a stall and seemed content. The herd was free to roam — just not outdoors.

No more picturesque herd in the Night Pasture until just recently — during the last five years — as Phyllis fenced the field and began putting her milkers and heifers out there.

Until then, instead of grassland, the pasture became (some would say) 'cluttered' or perhaps 'piled high with junk' as an ever-expanding area behind the barns became home to rusting machines, an evolving collection of discarded and salvaged farm equipment, water tanks, tractors, cannibalized bulldozers, refrigerators, antique potato diggers and orchard spray rigs, and — of course — way too many common junk cars.

There are also storage trailers: truck boxes no longer road worthy. Even our now-deceased neighbor ('The Maiden Lady') Elizabeth's old house trailer is out there, smothered under saplings and brambles, protected by resident hornets. You might say it's hard to hide a house trailer, but Elizabeth's discarded home blends well with the debris.

Other space at the center of our farm is covered by the large hay and dairy barns, the all-season cow yard, a mechanic workshop, abandoned cider mill and a seldom-used sawmill that once housed two old airplanes: a Cub and an Aeronca, occasionally a Tailorcraft or Cessna tail-dragger.

The most recent addition to the center-farm scene is a concrete manure pit that Phyllis built with federal farm subsidy money. Her small herd never delivered enough manure to fill a corner, but the government was practically giving them away to reduce pollution. Our father was all for the concrete installation. He wanted it big. Joked that it would one day serve as a foundation for an apartment house.

Now that our sisters have evicted Phyllis and her cows, some are wondering if she will get into trouble for not using the manure pit. Our parents had to give her a lease in order to receive the pit money. The lease doesn't expire for another three years and now she is kicked off the farm.

She says she confronted our mother with the eviction notice signed by Susie. It told her to be gone by October 4 or a cattle dealer would deliver the herd for slaughter on October 5.

"'Mommy', I said. 'Read this. It seems kinda mean. Do my sisters have more money than heart? Why can't I keep the cows at the farm where they will be safe and comfortable? I have a lease, you know. How can they kick me out?'"

Our mother quickly scrawled an amendment to the notice, stipulating that the heifers could stay. That's what it said: "The heifers can stay!" signed "Louise P. Harvey." Then, according to Phyllis, Mom ran out of the milk room crying.

Two hours later, our mother returned to tell Phyllis she didn't mean it and needed to cross off her note. She said the lease had been broken because of the words 'harmonious enjoyment of the property'.

Phyllis said: "She consulted with Erica ... APPARENTLY! And was told that the lease is broken by an absence of harmonious enjoyment."

Our mother told Phyllis: "Look at us. We're not harmonious."

Phyllis responded, also in tears: "You're not the only one crying around here. See! I'm suffering, too!"

These women hadn't cried together since 1987 when the herd left the first time during the federal Whole Herd Termination Program.

Back row: Albert, Dad, Matthew, Susie. **Front**: Phyllis, Mom w/Erica, Cindy, Beth, Me (Whoever said I don't look up to Dad? Was my hair REALLY that dark?!) When the photographer came, he asked: "Where are the children for the photo?" He didn't think we were properly dressed, (and I can't fathom why he sold us a picture with me and Erica looking in different directions).

Birthdates: Albert Adams 10/10/55, Daniel Webster 8/8/20, Matthew James 5/22/51, Susan Lois 2/18/50, Phyllis June 8/30/52, Carolyn Louise Peterson Harvey 7/9/25, Erica Alice 7/14/63, Cynthia Ellen 2/16/57, Martha Beth 5/19/62, Carol Ann 10/26/47.

Grandchildren: Albert: Nathan, Ned (Edward Nathaniel), Jared; Matthew: Analesa, Sharon, Danny; Erica: Martel stepchildren: Chris, Jonathan, Tara; Cindy: Abbie (Abigale Marie Harvey McPhee); Beth: Johnathan, Christa; Carol: Otis, Amos.

Historically speaking ...

9. N.H. DAIRY INDUSTRY - COWS

Before the American Revolution in 1776, life was un-evolved for the Colonist cow. Europeans brought the first bovines to Jamestown's Virginia Company in 1610. Then, it has been said, the Pilgrims never would have survived a second winter without a shipment of Red Devons from England in 1623. These early settlers could not have imagined that their thrifty foraging beasts would evolve into the designer species of today's high-production dairy herds.

Visions of Virginian village herds please me as much as imagining what it would have been like to tend the family cow in Boston. I came close when I was lucky enough to show my Guernsey heifer 'Bonnie' on Boston Common back when 4-H was an important part of any farm kid's life. Bostonian family cows grazed the 75-acre urban park until they were outlawed in 1830.

Can you imagine how brave it was for our rotund 4-H leader, Howard Allen, to haul our club's dozen cows into the heart of Boston's densely packed center?! I think it was called 'Cows on The Common' where maybe a thousand critters were displayed by area 4-Hers. We paraded and judged them as the Massachusetts and New Hampshire dairy princesses mingled with the crowd. It was part of a campaign to promote milk consumption.

In my mind, there is nothing (almost ... well, hardly anything) more beautiful than a herd of cattle grazing in a lush pasture on a sunny day. I regret that we seldom see such sights anymore in the United States, with or without grass and sunshine — though large majestic herds are continually displayed throughout New Zealand – black, white, brown and red cows against backdrop of green. If it's not thousands of sheep, it's hundreds of cows. Gorgeous rural scenery!

New Zealand cows live outdoors indulging their natural foraging instinct. They eat grass and walk several miles each day to and from the paddocks. It keeps them fit.

Herd health is as important to Kiwi dairymen as milk production. Their cows are not as bow-legged as American milkers because their udders are not so freakishly large. Special attention is paid to New Zealand cows' udders and hoofs. Ailing milkers are drafted into a separate herd pastured nearby for special care. Several times a year, a bovine farrier arrives to address foot problems. Similar to America's old-time cattle, New Zealand breeds are alert and fit with shiny coats to suit the changing seasons.

Until 1963, when Harvey Farm went to loose housing, our herd spent most of its time outdoors, restricted to the barn only during winter's coldest months.

At present, over 70 percent of America's dairy cows spend their entire lives on concrete in 'confinement' facilities where they never see, let alone taste, a fresh blade of grass. Instead, they are fed concentrated blends of fodder. By many standards, pastured cows are healthier. They do not ingest antibiotics that feedlots are forced to administer in order to prevent the acidosis that occurs when cows are fed grain.

That's what strikes me so strange. If a cow was built to graze and ruminate fresh vegetation in order to live and produce milk for its calf, how does the animal manage to digest rations of fermented balage, corn silage and hay, if it's lucky — and brewer's slop, bakery waste, citrus peel, palm husk, cottonseed, fish meal, peanut skins and ground-up feathers if less fortunate? Today's cows are sometimes forced to turn cannibal when fed butchery byproducts such as blood meal, tallow, urea and rendered fat.

Is that any way to treat the 'Mothers of Mankind'? If a Pilgrim cow ever dreamed it would be forced to eat FEATHER MEAL surely its milk would have curdled! Even the U.S.D.A. Cooperative Extension Service concedes that feathers might have 'low palatability' and must be introduced to the ration 'slowly' over a two-week period. These same federal farm advisors list two benefits of feeding slaughterhouse byproducts to cows: 1) Lower feed costs, 2) Disposal of byproducts in an ecologically sound manner.

Blood, fat, piss and feathers fed to herbivorous ruminating animals who would rather graze fresh grass!

Plus, it's expensive. New England does not grow the grains for pelleted bulk cow feed. This tonnage has to be hauled from the Midwest. Transportation costs increase the already high price of machinery-intensive production boosted through fertilizers and pesticides. Further ingredients in dairy feed include: corn meal, wheat middlings, chicken manure, almond hulls, apple pomace, ground barley, cull beans, beet pulp, waste chocolate, lupine and sunflower seeds, crimped oats and peanut meal, rye grain, canola and soybean meal, whey and mold inhibitors.

Still, this is better than the American disgrace of "swill milk" during the 1800s and into the 1900s. "Distillery dairies" supplied swill milk to the country's most populous cities — New York, Baltimore, Chicago, Milwaukee, Cincinnati, Philadelphia and Boston. Herds of malnourished cows were penned within filthy whiskey and beer processing plants and fed acidic hot spent grain slop until they died with ulcerated udders, lame, toothless, lacking tails (and the milk wasn't very good either).

Typically, drinking water was withheld to make the animals dependent on the slop for hydration. New Hampshire Extension Service literature stipulates that Boston's distillery herds were fed "brewer's grain," rather than the "slop" or "swill" of other cities. Even so, 19th century Boston city milk is described as "very thin and a pale bluish color, adulterated by adding starch, sugar, flour, plaster of Paris, eggs, etc. This milk was blamed for causing many fatal diseases in young children." (The History and Economics of the New Hampshire Dairy Industry, page 69, Cooperative Extension Service, 2007.)

Apparently, this is how it went in urban areas for nearly a hundred years before refrigerated transport made it possible to deliver abundant fresh food from distant rural areas.

Life expectancy for a Brooklyn distillery cow was 18 months. (Hard to believe, but it's in the book. They must not have bothered breeding for a second lactation. It was before artificial insemination, so it would've been difficult to manage bulls and monitor the cows' breeding cycles – especially with containment and overall poor conditions. So it was (probably) just easier to slaughter and eat the animals as they dried up. Dual-use livestock, you know.

Even now, an American milker expects to be shipped for slaughter before it reaches the age of five. In New Zealand, most herds have cows proudly producing as seven- to nine-year-olds. My Kiwi dairy boss says that's when cows reach their peak and he'd love to have a whole herd of seven-year-olds. But, of course, cattle CAN live to be 20 or longer if not stressed to deliver a calf annually and not milked nine or ten months each year.

Through genetic breeding, today's dairy cows produce 10-20 times more than a calf could drink. Industrial milk production pushes limits of creature comfort. Much criticism has been directed toward 'factory farms', which, I suppose, is a label appropriate for New Zealand's large dairy operations.

Even so, it is hard for me to see much of a downside for the New Zealand cow. I mean, if she has been bred to produce milk and doomed to waddle with a contrived mammary system, isn't that offset by a fairly pleasant lifestyle where she eats well and gets plenty of exercise?

Indeed, there was recent uproar among farmers and residents of the South Island's 'Canterbury' district over a proposal to build 'containment' facilities for 3,000 dairy cows. It made national headlines with people saying there is no way New Zealand cattle would be treated like American herds. They said it wouldn't be right to keep cows indoors and feed them mixed rations. Opponents promptly won and no building permits were issued.

I've worked mostly with black-and-white cows. Conveniently, because of their higher milk production, black-and-whites far outnumber the other breeds (like Jerseys, Ayrshires and beef-dairy crosses). So it's Holsteins and Friesians that I've experienced in the United States and New Zealand. To clarify, in Kiwiland, they have 'Friesians'," which are smaller tougher versions of America's trophy bimbo 'Holsteins'. Both breeds have black-and-white spots, sometimes brown from cross breeding.

New Zealand cows I've known seem content to live outdoors – especially where I work -- on the bottom of South Island. Neither too hot nor too cold, the area's maritime conditions are cooler than on the North Island. Also, there are no insects to torment the cattle – not like on the North Island (and the United States) where flies are annoying.

And what milker wouldn't enjoy free access to fresh grass following each 15-minute session in the shed? Herd managers strive to minimize herd time on concrete while waiting to be milked. After being collected for milking, cows enter the shed voluntarily. First ones through return immediately to grazing; last ones may remain on concrete for two hours. Rotating through 40 paddocks, grass grows for about 20 days between half-day grazing. The cows pass through the milking shed anticipating ample feed and refreshing time in the field.

Back on the farm in New Hampshire, it was a thrill to watch the cows race around and bellow as they were released into spring-fresh pasture. First day out of the barn after the long winter was an event where the neighbors were notified so they could join in the celebration. The cows were so happy! But that was just once a year and when gates were first opened to the several other grazing areas.

In New Zealand, through rotational grazing, the cows have the first-day romping experience twice a day.

And, regarding the milking itself, the cows don't seem to mind. I continually appreciate the general contentment within the large herds. Among 400 cows, milking after milking, very few of the cups are kicked off and there is no reason to moan about bad behavior.

Milking staff watch for health problems so the manager can draft and promptly treat the symptoms. If a cow isn't feeling well — suffering from sore hoofs or the onset of an infection called 'mastitis' in their udders or if they have stepped on one of their teats – well, I say their 'stompiness' is excused.

They (nearly all of the New Zealand cows I've dealt with), willingly 'let down' their milk. They seem to enjoy release of pressure within their engorged mammary glands. If tail-wagging is any indication, I'm here to report that cows often wag like happy dogs while letting down into the vacuum cups. And (from what I've seen), cows seem happiest in rotary sheds where each milker takes an eight-minute ride in its own private stall. They don't have to touch each other and are minimally handled by humans.

Back in Epping, with the small indulged herd, the milking process was regularly punctuated by kicks and hollers – way more commotion than with the Southland cows. I have a theory about this, and here it is: **Friendly cows kick milking machines off more frequently than normal cows.** I think they crave (even bad) attention – same as some people. Not much of a theory, but it's the only one I have and, hopefully, worth mentioning.

In the U.S. (as far as I know) herds are bred to produce milk year-round with cows calving in every month, but, as

Nature would have it, there is a bounty of spring calves resulting in seasonal gluts of fresh milk.

It's different in New Zealand where cows are 'dried off' around the autumn month of May. Dairymen shut down production for the winter. Except for 'town supply'/non-export products, NO milk is produced. No tanker trucks are seen on the roads. No bulk processing at the plant where they have already stockpiled concentrated milk solids for export.

To my continual amazement, the cows are all bred to calve during a two-and-a-half-month window spanning the springtime months of August, September and October. This is possible through computerized breeding programs putting entire herds 'in calf' through artificial insemination and 'clean up' bulls to end the breeding season.

From May into August and September, dairy herds are in the final months of pregnancy. The farmers are kept busy monitoring the herd. Their cows eat grass and silage, hay and 'balage' (hay that is not dried, but rather rolled into air-tight plastic wrap to ferment sweetly). Farmers often provide herds with fresh 'swedes' (rutabagas or Swedish turnips), foraged and uprooted where grown in the fields.

How unimaginably pleasant it would have been to just shut down Applehurst dairy production for the winter!

My current 400-head herd in Riverton NZ reporting for 3 p.m. milking. They are the 'Z' coming across paddocks viewed through new 'herd home' wintering shed used during calving season. When taking 'The Long Walk' from furthest paddock, the cows form a solid black-and-white line half a mile long (if that's what 900 meters equals).

Happy Hoser in Riverton, South Island. In Epping, we scraped cowyard with shovel and used a garden hose, rather than fire-brigade high-pressure hose common on New Zealand dairy farms. Back home, if we wanted more pressure we held our finger over the end of the hose.

1940: 'Farming and Flying in New Hampshire' - Dad written up as progressive young farmer with "24 4-H projects, 50 acres orchard, 3 acres vegetables." (Reverse of clipping: 'A Time-Table of Dictatorship'; seems some things never change.)

Excerpt: '…Dan Harvey for three years has been leader of the Pawtuckaway 4-H Club. Last spring, when Mr. Harvey injured his knee, Dan had an opportunity to show what he could do on the farm, a responsibility which he met in the usual 4-H manner …While farming is the main interest of this 4-H club member, flying is a close second. He holds a pilot's license, with 15 flying hours to his credit. Looking ahead, Dan plans to build up the fertility on Applehurst Farm, with use of the most modern methods, and at the same time reserving one plot as a landing field. Early last spring he flew to East Otisfield, Maine, to purchase a cider mill in order to have it ready for operation on the farm this fall.'

The way we are...

10. RETURN TO THE NIGHT PASTURE

The Sisters' Refrain (please) ...

We know that some people say
Losing money is the Harvey way.
You can call it tradition
But it still lacks ambition.
'Cause taxes are here to stay.

***

Of all the things under the sun
We forget about when we are done
Is where the hatchet is buried.
But why are we worried?
We can always buy a new one.

Certainly, my sister Phyllis's small herd received better care than most contemporary cows — though she did serve up 'slop', er 'brewer's grain', from the Budweiser plant in Merrimack. Her 30 cows and young stock were pampered and allowed to graze the Night Pasture, which was a welcome return to the old days.

People commented on how much they enjoyed seeing cows in the field. The herd was so 'pastoral' so 'picturesque', a rare sight since herds have been kept out of sight for so long. Our 150-head herd terminated in 1987 had not been outdoors for twenty years. The cows had enjoyed amenities provided by 60-years-worth-of-concrete featuring feed alley, comfort stalls and milking parlor.

Not just bucolic, Phyllis's milkers were healthy and enjoyed the freedom of grazing. Still, her pasture management could not compare with New Zealand's 'rotation' methods. My sister did not maintain the Night Pasture for maximum freshness and nutrition. She could not afford to install water troughs, movable fencing, and manure irrigation.

My Kiwi herd owners are as attentive to the condition of their paddocks as to the well-being of the cows. They say good grass makes good milk. New Zealand pasture offers abundant grass through highly-developed drainage and manure irrigation infrastructure.

Weeds are controlled and mature grass is continually mowed to provide the herd with rich grazing after each milking. It's not as simple as just slapping the cows on the bum and saying: "Go for it!"

Not coming anywhere close to that level of forage, even Phyllis's 20 cows spent most of their time clomping around indoors, dependent on mixed rations served along the feed alley designed to feed 100.

New Zealand dairy owners just shake their heads in disbelief when I tell them about America's reluctance to provide cows with pasture. They say cows SHOULD eat grass. I can't explain it to them, except to note that New England has deep-freezing winter and New Zealand does not. Maybe we just can't grow grass in New Hampshire (though I'd say we have a better crack at it than Western factory farms with desert conditions).

There is definitely a difference in "dairy culture" between what I know in the United States and what I've been exposed to in New Zealand. Several of my Kiwi employers have toured Midwestern United States, where they judged our dairy operations to be "disgusting and backward." They are critical of American dairy farms with feedlots and giant cows with uncomfortably-large udders. They think American dairy farmers are driven too hard by capitalism to produce milk at any cost to the cows and cropland.

'Big time' dairymen I know in the United States seem perpetually disgruntled about low milk prices. The government has always monitored and manipulated agricultural markets. Yet, price supports and dairy surpluses provide little benefit to American farmers – especially the small ones still trying to play ball with the big guys.

Ironically, New Zealand's farmers cheerfully benefit from exporting to Asia, Europe, Mexico and India. Through an unapologetic socialist economy, New Zealand's unregulated commodity prices allow farmers to take advantage of expanding international markets. Food prices are high in New Zealand. Domestic consumption is not subsidized. Farmers are paid well for their product and can't deliver enough of it to meet the ever-increasing export demand.

Back in New Hampshire, it didn't seem to matter how hard or how well we worked. The milk check did not cover expenses. It's not as if our parents didn't try to make a go of it.

Take note! My parents installed a NEW ZEALAND design herringbone milking parlor in 1963, one of the first, if not the very first in the state! Dad paid attention and continually enrolled in government programs. He signed up for any help that came along, whether for erosion control or fertilizer and seed. Our farm pond was engineered and built with federal assistance and miles of government-subsidized drainage tile buried under cropland.

My parents were progressive and brave enough to borrow money and make improvements. They believed our government and the milk company when they claimed "bigger would be better."

We expanded to be modern and should've been contenders!

But it didn't work out that way.

Uncles Bill Woodward of Durham and Philip Peterson of Fremont, both deceased.

Scenery long-gone: Can you see the cows grazing beyond pond by sawmill? See that apple tree by the pond? That's where the ill-fated 'Sadie Lee' stood when she was struck by lightning.

Phyllis in final days of milking at the farm. She is showing Cousin Tyler Young that she uses his product: Absorbine.

The way we were ...

11. JOHNNY YUMA

I bought a CD the other day and didn't realize until I heard it that Johnny Cash sang the theme song to one of my old television shows: "THE REBEL, starring Nick Adams as Johnny Yuma!" That's the way the announcer said it. Susie, Matthew, Phyllis, Albert, Cindy, and I claimed our places in front of the tube as the 14-inch screen filled with black-and-white images of a handsome blond boy shooting a sawed-off shotgun.

Half an hour later, we'd be holding hands — well, I, at least, would be holding Cindy's hand because she was only four — while walking through darkness to 'The Little House'.

Funny the way a song can take you back. "*Johnny Yuma was a rebel. He roamed through the West. Did Johnny Yuma, The Rebel. He wandered alone ...*" I used to sing it.

Twelve years old, barefoot under a drippingly-humid starlit sky, leading my little brothers and sisters past our stinky septic pool. We felt our way through the hayfield to the pond. No flashlights, but there was a path to follow, the width of the mower. I cut the walkway every week to keep it short and soft. One step off the path and we'd know it; the hay stubble was so coarse in comparison. Straight path to the pond, then sharp left along water's edge to the door of The Little House.

Our pace and mood during the 5-minute trek depended on the tenor of "The Rebel" and the phase of the moon. We rushed if Johnny had had the chip knocked off his shoulder too many times. It made me anxious if the show ended in distress. The Confederate soldier fought if he had to and fell in love countless times, always riding off into the sunset to find himself, alone. Belligerence or romance influenced our footsteps and stargazing along the path.

If the moon was new or if it was raining, our walk could be balky and treacherous — especially if we missed the turn and slipped into the pond.

Often, we were tired at that hour and on the verge of tears. Not me, of course, but the littler kids. Mom would say,

"Oh, he's fallen asleep" or "She's too cranky. You'd better leave (whichever baby) here." But we thought we were "The Six Musketeers" (another TV show: 'The Three Musketeers') and pretty determined to sleep outdoors. It was "one for all; and all for one." We scurried into the darkness before Mom could snag our weakest.

In 1958, Uncle John (who wasn't really our uncle) used The Little House to teach us the basics of carpentry — the names of various construction components (sill, plate, header, joist, rafter, etc.), and techniques (toenail, sister, level, plumb, shim, etc.) — and how to use the tools to build according to 'Code'. Uncle John was a real carpenter who had built where there were actual building codes, but Epping had no Code ... and never would if Dad had his way.

The pond-side cabin was Matthew's idea. Thanks to my brother and Uncle John's indulgence, we kids slept by the pond every summer's night until school resumed in early September.

At age eight, Matthew asked Uncle John how to design this "little house" that he intended to build by himself. He had watched and asked tons of questions as Uncle John built and remodeled our cow barns. The 'New Barn' was a 50-cow tie-up, a never-ending project that John was kind enough to keep plugging away at on weekends for years, even after the herd moved in.

Our barns were the best around, thanks to Uncle John helping Dad and sometimes Uncle Eddie with major alterations, like lowering floors for more headroom and installing stanchions, drinking cups and gutter cleaners. Those were the good old days when property owners could do whatever they wanted and didn't have to worry about building inspectors.

The Little House was John's recreation. He seemed to enjoy watching us kids mark boards, straighten bent nails and bind the saw with crooked cuts. He followed Matthew's crude blueprint, adjusting measurements to make it work.

We called him "Uncle John" because that's what he wanted, though he had real nephews and nieces in Osage, Iowa, to where he eventually returned and has now passed on. He was a minister who married and buried friends and relatives requiring his services long after he moved back to Iowa. I don't know how often he came back to his Jonesport

(Maine) and Brentwood Baptist (New Hampshire) churches, but it was plenty. People phoned in tears and he came back.

John (Gilbert) was married and divorced before I was born. The only thing I knew about 'divorce', was don't ever mention it. Walking away from marriage wasn't approved back then, especially for ministers. So our uncle spent what seemed to be a long time 'away' being a pastor in Jonesport and Jonesboro, Maine, where he met his second wife, Agnes. They married in 1970, surprising everyone when our new 'Aunt Agnes' abruptly left her hometown and moved (at age 48) to live with John in Iowa, where she remains as his widow.

Her Jonesboro friends still marvel: "Who would have thought Agnes Lincoln would end up in Osage, Iowa?"

When we grew up, Matthew's 'Little House' was hauled to Cindy's for her pigs. Not sure whether that was before or after the pondside apple and nut trees were cut down, but I DO remember there being apple trees and a tall hickory by the cabin. Several cows were struck by lightning, back then, while seeking shelter under those trees.

Just now, during the past two summers, the original Little House has been replaced by 'The Pond House', built as another learning exercise by our son Amos and his cousins, our teen-age nieces, Christa and Abbie (Beth and Cindy's daughters).

Anyway... back in the 1950s, we Harvey children did not always rush to sleep in the poorly-ventilated, often mosquito-infested cabin. Appreciating safety in our numbers, if it was a quiet moonlit night, we lingered in the field between the farmhouse and pond. We gawked at stars and caught 'lightning bugs' to put into a jar by our beds.

Evening air stirred with the nocturnal beat of a hundred bullfrogs thrumming from the basin of the pond, so much more dynamic than the stagnant residual heat that awaited us inside the Little House. Our hayfield was cut twice each summer. Depending on the hay, whether it was emerging, maturing or being harvested, the crop emanated aromas of clover, alfalfa and timothy. Dew was usually heavy around the pond as cold air settled from the nearby hill. This saturating moisture freshened fragrance of rich grass and lifted an organic aroma from the aquatic world around the croaking pond.

Whip-poor-wills were alive and well in southern New Hampshire back then, though these birds are seldom heard today. At dusk, they called: "WHIP-poor-will! WHIP-poor-will!!" Deep-throated bull frogs bellowed from cattails growing in the shallow end of the pond: "Br-wum! Br-wum! Jug-o-rum!" Their rolling drone rose and fell as if one Big Brutha' Frog directed their crescendos and fades. In early summer, peepers made a racket and tree toads tweeted like referees at a basketball game. On through August, crickets and cicadas revved up for that dreaded first day of school.

It's a lot of noise to compete with while you're wailing about Johnny Yuma's gripes as a young soldier who was *"panther quick and leather tough and figured that he'd been pushed enough! ... The Rebel!!"* Nick Adams was SO cute in his little Confederate hat! I especially admired him for taking time to write in his journal. You just don't see that much good penmanship on TV anymore! I think his gun was one of his problems, though. He seemed to regret sawing it off too short. You had to feel sorry for Johnny. Not only had he ruined his shotgun and couldn't afford another one, but he'd also lost the Civil War.

Whoever reached 'The Little House' first turned on a big lantern, our interior lighting. It ran off a car battery, so lasted the whole summer. By the time everyone was inside so were dozens of bloodthirsty mosquitoes. We picked most of them up while walking the path beside the pond, so made a rule to RUN QUICK! The last distance and shake them before entering the house.

Four small half-screen windows were not enough to ventilate the 12x12-foot structure. Our north wall had no window at all because the bunk bed spanned all but three feet of it. Summer nights were usually windless and stuffy, but we didn't complain. There was no one to complain to.

All six of us slept in the almost-queen-size bunks — three on top; three on the bottom. Later, I think it was boys on top; girls on the bottom. And, whenever friends slept over — always girl friends; Matthew and Albert never had anyone over — the boys were happy enough to sleep outside in a tent. No matter how we sliced it, our summer nights lacked ventilation and featured talking long into the night.

We loved our Little House.

You call this 'skating'?! Erica, Ned, Albert, Jared and Nathan on pond. We enjoyed the pond year-round: swimming in summer, skating when frozen. One year, Susie won First Prize in a fishing contest. Epping icon Jack Sharkey (a world champion boxer who fought dirty) complained that it wasn't fair because Susie had raised the fish in our pond. Susie said: "Tough!"

The way we are ...

12. WRITING ON THE WALL

The Sisters' Refrain (please) ...

> ***'Illegitimi non carborundum'***
> ***Was always Dan Harvey's profundum.***
> ***The boys are cunning***
> ***To keep machinery running.***
> ***You just gotta know how to fund 'em.***

Dad was 27 when he inherited Applehurst Farm. Aunts Edna and Aggie were 26 and 25, respectively. I've always wondered how that male-takes-all transfer was determined. Of course, it was the way, back then, to have girls marry and leave town. My aunts received 'provisions' of $1,000 each to save breaking up the thousand or so acres. Grammy Harvey lived another 17 years and didn't seem vested in the farm.

I don't know how it worked. But certainly it isn't the way it's going today among my five sisters and two brothers, where Matthew and Albert enjoy no special privileges, no more than the rest of us. Some might even say they've been passed over.

Matthew, in particular, was discounted when the herd was terminated in 1986/87. That's when my brother clearly saw the writing on the wall. We had all turned on him, deciding to throw in the "slobber rag," as we referred to the towel for washing udders. We were getting rid of the cows. My brother told Dad he would do one final harvest, then accept an offer to do construction work.

Matthew was 35 at the time. Up to that point, he had devoted his youth to working on the farm. Before and after school, he did 'chores', declining to participate in extracurricular activities like basketball and baseball. My five younger sisters – not Susie so much, but the rest of them — played sports. They were all talented in basketball and softball. They won trophies. Matthew and I did not play sports and I don't recall Albert being much of a dribbler. If we were athletic, we never proved it, and we certainly don't have trophies.

Part One - Terminated

I remember deciding that I didn't want to do anything extracurricular. It seemed an odd activity – sports — hanging around after school to play games and then traveling in a bus to compete with other schools at night. There was no Physical Education ('PhysEd') offered at any point during my schooling. Everyone was on their own when it came to exercise. By the time a teacher was paid a stipend to coach high school basketball, softball, volleyball and cheer-leading — all after school — it was too late for me to learn any of it. Teams were drawn from the kids who had already been playing together downtown. I was on a Math team once, but we were never celebrated by cheerleaders or rode in a bus to compete or snapped each other with towels or ate orange wedges during time-outs to charge past our opponents.

Strange, isn't it, that my team experience came on the farm doing chores and field work with siblings — not with peers at school? Sour grapes. I didn't want to play basketball anyway. My excuse was raising a couple dozen calves and tending four horses. I'm not sure what Matthew's excuse was, but he certainly fed out many tons of silage to the cows after school and played hooky a few days to plant and harvest.

My brother probably assumed he would take over the farm from our parents, same as the previous five generations. Indeed, early on, Mom told him he would one day own the pond-side of the road, across from the farmhouse and barns. But that's one of the things we've forgotten — the reason he built a house over there, which now provides rental income for our parents.

And what about Albert? With Matthew and three sisters ahead of him, what future did he have in farm management? Straight out of high school, he made his own way as a family man and mechanic.

Matthew focused on the farm and devoted more time to the effort. Always good with crops, he ran equipment to plant, cultivate and harvest hay and corn for silage. Plus many of the infrastructure improvements — the cow yard, feed alley, trench silo and a monitored feed system — were Matthew's initiative.

Dad continued to call the — mostly political — shots and spent more money than anyone else, sometimes over family objections (as for more land or more used equipment or perhaps an eleventh spare-parts Caterpillar D-7 bulldozer). Matthew's name was not on the checking account and, as far as I know, he did not sit down with our parents to sort through tax bills or discuss farm operating expenses. If a management strategy existed, I don't know what it was. Matthew had a wife and three kids when he stopped working on the farm.

One of Dad's political peeves cost us extra tax money, paid to the Towns of Epping, Nottingham and Lee.

'Current-Use Taxation' was adopted by statewide referendum in 1968.

"YES!" Dad exclaimed. "Land SHOULD be taxed for its current use — not for its potential use. PERIOD!"

He said intent of the referendum was good. Yet he refused to enroll the farm for current-use assessment. Our father could not abide amendments to the referendum. He said the attachments were "against his principles." He didn't like the way the New Hampshire legislature imposed a 10 percent conversion penalty when property eventually changed use. Dad said the fee would do nothing to slow development and would be paid by first-time homeowners who did not deserve to be punished. He said people who already owned homes should not be 'privileged'.

I was a good daughter, supporting Dad during many current use discussions. I was proud of him for standing firm. Still, these guiding principles for politics had little to do with dairy production. Dad's stance caused distress while diverting much of the milk check toward land taxes. It also showed that one person COULD make a difference.

Facts aside, what would've happened if we didn't kill all of the cows during the federal buy-out?

What if we had resisted the government offer to pay off all debt if only we'd stop milking cows? How could we have decided to continue juggling all the balls while our paternal ringmaster kept borrowing money to pay taxes while we worked for peanuts?

I admit it. I was for terminating the herd. Definite no brainer. A choice between all the energy and resources going for no income. Or pay off the long- and short-term loans and give it a rest. It seemed that nothing short of a federal order and hand-out would make us stop milking cows. Finances and management had become too complicated and stressful.

Even the Department of Agriculture said farming wasn't worth it, urging dairymen to cease the futile effort.

But what if Matthew had said: "NO! We're NOT pulling the pin!"?

Wouldn't THAT have been some sort of shake up?!

To this day, Dad refers to himself as 'The King' and says he still runs the farm, though inertia is a key element of his management style. He has explained 'inertia' to me a hundred times: "A body in motion tends to remain in motion until outside forces are applied, while a body at rest tends to remain at rest." Dad is not one to accept or promote change. Status quo suits him fine: "Just keep a-doin'!"

After all, as Mel Brooks once said: "It's GOOD to be King!"

Twenty-two years ago, my father was even less willing to assign responsibility for running the farm to Matthew, as his own father had done for him when he was 27.(Of course, my grandfather's sudden death pushed the agenda.) Today, on the cusp of 90, Dad isn't ready to step down or surrender his beloved "rubber checkbook." He reigns as King.

My executrix sister Sue may wear a golden bonnet, but it's still not the crown.

So why speculate on what could've happened in 1987 ... because it really COULDN'T have. The cows were killed and there's no changing that. It might've been Matthew, instead of Phyllis, who resumed milking in 1994. He could be the one all worn out today, rather than The Fair Flower. Our brother might've enjoyed more Harvey support. But I doubt it. As it has turned out, Matthew's family is thriving without depending on the farm for groceries and happiness.

This is what we were driving when Marjorie Armstrong Coulter shook her martini mixer at us in Chapter One. Picture shows Matthew on 1940 John Deere 'A' trying to reach pedals.

My brother Matthew, owner-operator trucking and stump recycling business, resides at 'The Crocker Place' with wife Judie and offspring: Analesa, Sharon and Daniel James.

Barn before recent renovation. Pole on left ridge also restored as artifact from Revolutionary War. Roof lines have been straightened, siding repaired and painted. When pasture replaced orchard, the old milk room was in this end of long-slope roof (shown in disrepair). Cows stood on plank floor, chained along wooden water trough straight in from milk room. After six cows, trough took 90-degree right turn to west to accommodate the ever-expanding herd.

Pumping diesel fuel from tank Up Back.

Party Barge at Camp Menatoma, Readfield, Maine, where our kids grew up with these classmates: Andy, Otis, Aaron, Melanie, Patrick and Amos

PART TWO

The Way We Were

13. MUTUAL BENEFIT THROUGH MARRIAGE

On the Peterson side of our family, Mom was the first-born, like me, an experimental start of a young family.

My Peterson grandparents were clear about their expectations and provided solid models for their children. No doubt, growing up as Mom did with straightforward guidelines, she had firm traction for getting ahead, whereas we, her eight offspring, have ventured forth on more slippery footing. We have been spared many of the rigid standards that shaped our mother.

Whether that helped or hindered us remains debatable.

Certainly, through Mom, we eight benefitted from Petersons' fortitude and perseverance. Unlike other mothers, Mom seemed confident in our ability to deliver. She is fearless and didn't fuss over our safety. Didn't scare us with warnings about pedophiles, knives or explosives. She gave us as much responsibility as we could handle and never assumed we'd fall short in doing our part to keep the house and barns in order. I believe that same expectation extended to our school work and beyond.

All eight of us sought Mom's approval. We "behaved." That's what Grammy Peterson frequently snapped: "BEHAVE!" and gave us a yank. Mom finessed the concept by believing in us, using psychological, rather than smacking, guidance.

Even today, as adults, we eight 'kids' elbow each other aside to please Mom because she has earned our devotion. Even when we screw up, she encourages us to be individuals and follow our dreams if we happen to have any.

When they married, Mom was gracefully influenced by Dad's more playful approach to life. She shook Petersons' flinty skepticism and caution to accommodate the relationship that has emanated love and mutual respect for over 60 years.

Dad wasn't raised same as Mom. Three generations removed from immigration, he inherited the yoke of Yankee independence. Rather than accommodate, Dad was taught to demand freedom and question authority. His forebears helped settle the towns where we still own property: Epping and Nottingham. His father and grandfather moderated town meetings and, as near as I can tell, they were well respected within the community. Epping's public library bears the name of a very-great-uncle who donated money: 'Harvey-Mitchell Memorial Library'.

Dad has two older (now deceased) half-brothers who visited regularly from where they settled in northern Massachusetts, plus two younger sisters, Edna and Agnes. My grandfather was 50 when Dad was born and my grandmother had already raised a separate family. The birth of young Daniel was greeted with joy and the confidence of life experience.

Dad was taught to involve the farm as a vital component of Epping's community life, its civics. There was no question that he would become a player in town. This was different from the Petersons who tended to keep their heads down and kept working, seldom seen on Main Street.

While some of my Harvey ancestors had certifiable psychiatric problems, others (well, actually, some of the same ones) were recognized folk artists, craftsmen and writers. Dad always wanted to publish his grandfather's poetry, but Mom referred to the often-misplaced artifacts and manuscripts as "piles." Dad did not see why she had to clean things up before he was finished making a mess. He is comfortable with clutter; she is not.

Dad grew to appreciate debate and witty verbal exchange, while Mom has been known to say "Bah!"(a Peterson term) and discount people who don't agree with her or, more likely, those who don't work hard enough to please. She has not totally shed her Peterson-Cole ways.

It was important for Petersons to milk the cows and do chores on time, at 5 a.m. and 3 p.m., same time every day, 365 days in a row ... then start over. As first-generation immigrants, (Petersons from Sweden; Coles from Canada) they toiled with a vengeance for everything they had.

All I remember about my great-grandmother Cole was that she wore black and sat in a rocking chair on the way to the outhouse. Every time I passed her she said: "Use just three squares of toilet paper." I think the stuff had just been invented and three squares wasn't much … and not much of a memory either. But better than nothing, which is what my younger relatives recall.

There's a story about my great-grandfather Henry Cole (Grammy Eunice Peterson's father) losing his temper in 1936. On Christmas Eve, when Aunt Ann was barely four months old, Henry lost it and kicked his son-in-law (my mild-mannered Grampy Albert Peterson) out of the house into the frigid winter's night.

"They say Grampy (Henry) never expected Momma to leave, too, with us kids," Aunt Ann said. "But, of course, we all went. They loaded our Christmas presents and roped my baby carriage to the top of the car."

After a few days, work forced the family back to the farm. My great-grandfather persuaded his son-in-law to return because the chores were too much for him. He said he'd do anything to get the family back.

And that, My Children, is how the Albert Petersons acquired Brookvale Farm with a vengeance.

When my mother (Louise) was in high school, Grampy built her a commercial chicken house. She raised and sold eggs from 1,500 hens. She received day-old chicks at the Epping train station and mothered them around brooder stoves in the top of her new chicken house. Until the chicks grew big enough for her not to worry, Mom slept out there with them. She also raised Collies and sold pups. Aunt Ann remembers 'Vicky' and 'Hones' as Mom's favorites.

Meanwhile, the nearby Harveys seldom worked that hard or did things on time. Dad says my grandmother (Alice Webster Harvey) refused to lock herself into a meal schedule.

"If we were hungry, she'd tell us to go get something to eat." Apparently, it was no big deal.

Dad was not expected to do chores on time or even every day. He seemed to miss out on guidance through structure, guilt and intimidation. Among the Harveys, punctuality was not something to be proud of.

"There was time to do whatever we needed to do," Dad said.

And, instead of an egg factory, his father gave him a play thing: an airplane just "to keep the boy on the farm." He was exempt from fighting in World War II because he was an only son and had to continue farming. With the Piper Cub, he joined the domestic Civil Air Patrol and flew out of nearby Hampton airport.

Apparently, the airplane worked as an anchor to the farm. Dad is content with not straying far from Epping. Mom is the one who longs to go places and see new things. Dad gladly goes along with whatever she arranges, even if she tells him he has to stay home and "tend to things." He has always been able to "hold the fort" while entertaining himself.

My parents have traveled quite a bit — to Alaska and Hawaii several times, and back and forth to Iowa and Florida where they have friends.

Mom has been to Europe and elsewhere without Dad. I've always felt that she would've preferred not marrying and settling down so soon. At least that's what she told ME to avoid while encouraging my longing to be 'free' — unencumbered. She understood when I declined her offer of a portable sewing machine as a present when I graduated from the University of New Hampshire in 1970.

"No. It's too heavy," I said. So, while others received new cars, I got a two-piece bathing suit.

A favorite memory of my mother happened late one night, well into my adult years, when my friend Maureen and I approached the farmhouse after driving down from Maine. Just Maureen and me minus husbands and sons, arriving to spend the night. The outside light snapped on and my mother came out waving her arms while singing Aretha Franklin's "Freedom! Freedom!" She hugged us both. I flushed with pride, seeing her in such good form.

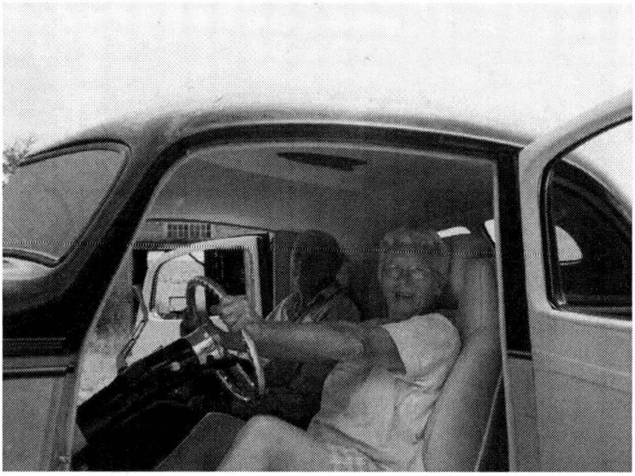

Mom at wheel of a developer's hot rod. She said he stopped coming around when he figured out we wouldn't sell him any land. "If we did, it would've looked as bad as Hetnars'," she said, referring to an abutting orchard now cut up into one-acre houselots.

Mom loves going to her friend Mary Ellen's oceanside cottage on Baileys Island, Maine.

The way we were ...

14. THE GOOD OLD BOYS

Note: I can't remember the name of Schofields' boat. But you can bet your boots it wasn't "Cindy-Lynne."

"I'm not sure we CAN launch the ass end first," said Arnold, rubbing his double chin as he studied the cabin cruiser cradled and chained to a trailer that Dad had borrowed for the day.

"How will we ever turn it around?" Arnold wanted to know.

Everyone's favorite mechanic, Arnold Dennett, had rebuilt and installed a flathead V-8 in our neighbor Parker Schofield's 24-foot-long homebuilt boat. The two of them — along with Dad and Noel Gosnell, (a young Floridian who had helped Parker with his garage project for the past year and a half — and Cass Adams, (our cousin who was a real boat builder) — all stood in the parking lot above the Newfields boat launch.

Tidal water rose imperceptibly in the Squamscott River as it flowed under the 'Singing Bridge'. Nancy Schofield and her daughter (my best friend) Cindy had picked up Mom and us kids to watch. Four of us then: Susie, Matthew, Baby Phyllis and me. I carried my Brownie box camera.

"How do we see this working?" Parker asked, quite willing to share responsibility for the launch. "I haven't thought past building the thing and didn't know for sure until this morning if we'd be able to use the trailer." His varnished plywood-and-mahogany boat sat tall, facing backwards, behind our 1940 Federal dump truck.

With high tide two hours away, the murky river water seemed incredibly low. It was WAY DOWN while the proud brass-clad bow of the "Cindy-Lynne" was WAY UP there in the parking lot. She perched high above the ground at the top of a steep ramp to the water.

Cass Adams knew about rigging and had loaded his 1938 Chevy ton-and-a-half with the essentials:

ropes, blocks-and-tackle, planks, rollers and two logs that he called 'shears'.

Dad and Arnold had been telling Cass about our neighbor's boat project for a couple years, but this was Dad's cousin's first look at it. He seemed to approve and Parker was grateful for the seaman's offer to help with the cabin cruiser's maiden voyage.

"We just need to hold steady, not rush it," Cass said. "We don't want to hurt this beauty before she gets her gunwales wet." He said we might have to let this tide go and wait until tomorrow if they couldn't set things right. He said it would be best to launch her stern first. They should try to turn the boat around.

Our cousin's ancestors, like ours, were early settlers with the advantage of claiming some of the area's finest real estate. Harveys built on Red Oak Hill. Adamses staked out land along Great Bay in Newmarket. Indeed, today's state-owned "Adams Point" was named after Cousin Cass's forebears, (including the U.S. President Adamses).

We used to go there all the time, every summer weekend, less often during winter, for picnics and to help Cass fill his barn with hay for the horse and a few cows.

I remember Cass's father, Captain Ed Adams, reclining as he smoked his pipe on one of their three divan sofa things in the dark smoky living room. This dingy room would have been more correctly called a 'den' since it resembled a cave. Innermost rooms were deprived of daylight because a wrap-around porch shaded three sides of the embellished Victorian-style house. And there were no lights. No electricity at Adams Point. No plumbing either. We kids were always asked to lug buckets of water from a spring beyond the barn.

Uncle Ed Adams's father Joseph ran the (now destroyed) house as a summer hotel in the early 1900s. It had six ornate dormers and a wide central staircase to accommodate guests on the second floor. A narrow passage off the dark kitchen led to two three-hole indoor outhouses — one for men; one for women. Each hole was a different size — small, medium and large.

For wiping, there were boxes of newspaper torn into squares.

Since Cass was 30 years older than my father, Dad was always the young guy giving his relatives a hand. The Adams women stayed in the kitchen, out of the way so the duffers could lay around smoking pipes and cigars by the glow of kerosene lamps. They frequently napped and I think both Captains Ed and Cass had back troubles. A polished nickel-trimmed parlor stove burned coal and made the men cough as they stoked it. They wore layers of gray woolen clothing and didn't have much to say to children.

In contrast, everything outside the gloomy den was radiant and magical. The exterior rooms and bay windows were filled with Adams Family folk art. The house was like a disorganized museum with oil paintings of scenery, horses and birds, mixed in with human heads carved from wood and decorated with hand-woven hats and scarves. One long-necked pine head wore a World War I helmet. Wooden duck decoys and other sorts of life-size birds, fish and turtles were scattered among antique furniture.

Nobody lived in this part of the house, not since it no longer served summer guests. There were the most amazing models of clipper ships and cargo boats detailed with tiny ropes, blocks-and-tackle, and little wooden sailors on deck. Also, handmade jewelry and collections of shells, stamps, buttons and dried pressed flowers.

There were several models of the Adams trademark: the sailing barge "gundalow." Gundalows are flat bottomed, shallow-draft vessels once common on local rivers and salt-water estuaries. These sturdy cargo carriers were common along the coast of New Hampshire's Great Bay and southern Maine. Gundalows were among the very first vessels built by early settlers. The last commercial gundalows disappeared from our waters in the first decades of the 20th century.

The Adams house collection was donated to Mystic Maritime Museum in Connecticut in 1962 after a long weekend of auctioning off the lesser collectibles.

The Adams women knew how fascinating these oddities were for children and opened the doors only if we were 'good'. Candy and ice cream were no match for the treat of being escorted through the house.

In the barn, Rhode Island Reds guarded the boat-building shop, as the hen coop was partitioned by mere netting. The workshop smelled of turpentine, shellac and horse manure, since it was over the stable, plus ammonia from chicken poop. The area appeared static — same timber lying in same places, week after week, year after year. Cass said it took time to find trees with just the right curves for boat building. Besides, there was his bad back.

Then, there was Adams Point itself, the land, with its peninsula of undulating wind-blown grass that extended into the bay with a shallow border of hickory, oak and maples along the mile of black shale cliff that divided hayfield from placid sea water.

The horse and cows grazed in a meadow with a meandering stream around to the west where Dad's artistic Aunt Avis rendered landscapes. Unlatching the weather-board and hammered-iron gate to the meadow was yet another Adams delight as it opened visually rewarding vistas.

At low tide, the dark sharp-shale beach crawled with horseshoe crabs, snails, and centipedes big enough to eat mice (not that they would; but they might!) We squinted through a crack in the door to Adams Tomb and saw caskets. That was pretty cool, as were the gazebo, floating dock and granite-and-wrought-iron benches placed at overlooks around 'The Point'. The Adamses appreciated scenery and did not neglect their art and crafts.

There was nothing better than playing on the gundalow. Cass had built it with his dad and they launched it together just before the old sea captain, my father's Uncle Ed Adams, died in 1950. 'The Driftwood' was the real deal, probably 50 feet long, moored by the house and floated only during the highest tides. It was a smaller version of 'The Fanny M', reputed to be the last working gundalow in the world, named after my Great Aunt Martha Frances Harvey.

(A 'Captain Edward H. Adams' replica was launched by Strawbery Banke craftsmen in 1982. Today, it is a popular vessel used for education along the Piscataqua, York and Exeter rivers.)

So that's where Cousin Cass came from and explains why Parker Schofield was so happy to have him at his little 'Cindy-Lynne' launch party.

"How, exactly, did you get her onto the flatbed?" Cass asked.

Parker gestured toward a bundle of rope in the back of our Federal and explained that, just yesterday, he and Noel had finished making a sling for the boat and a shallow cradle with come-alongs to fit snug between the wheels of the trailer.

"And, at six this morning I brought over the 'G' and we dangled her on," Dad said, leaving out the part about nearly causing a premature shipwreck when, mid-hoist, the John Deere's rear wheels began bouncing and the boat's stern twitched precariously

Neither did he mention that Mary (his friend Otis Goodrich's wife), objected to his borrowing the trailer. She said he would ruin the rear wheel bearings by submerging them in cold water after the heat of a 15-mile trip to Newfields. Dad didn't say anything about this because he knew Mary could be right and she was still mad at him for poking a hole in the radiator of the last piece of equipment he borrowed.

But why ruin a good time with full disclosure?

Dad was enjoying himself. There's nothing he likes better than a challenge, a conundrum just hopeless enough to require his out-of-the-box input. He said placing the boat on backwards was "a nice touch."

"If we're gonna try swingin' her 'round, we'd better do it right here before we back the rig down there," Arnold suggested. "Why the HELL did you put it on there backwards, anyway?!"

Dad shrugged and offered an excuse about the tractor not being counterbalanced well enough and there wasn't enough room in Schofields' driveway to maneuver the truck.

"Besides, it was just Parker, Noel and me this morning. We figured we'd wait until we had a full crew and all this assembled brain power," Dad challenged.

My father and Arnold were best friends. They shared many adventures together, not the least being the purchase of two Aeronca airplanes from their mutual friend Sammy Tecce, an aircraft engine mechanic. Dad bought a 'Defender'; Arnold got a Piper J5 "Cruiser," which he immediately crashed into power lines and said he'd never do that again. Dad was always saying what a "genius" Arnold was, for a man who hadn't gone past sixth grade. He bragged that Arnold could fix anything and claimed, "Even the Navy Yard comes to him for submarine parts!"

Of course, Arnold paid a price for Dan Harvey's love and respect. My father was forever "over at Arnold's" either "chewing the fat" or "holding Arnold's hand" while making repairs. Dad seemed to camp at the Newmarket shop and couldn't help mentioning each time he had to wait for Arnold to get out of bed.

Arnold's sister, Doris (I think), served the men many lunches consisting of beans and toast or Campbell's tomato soup and a cheese sandwich or Chef Boyardee ravioli. Then, waiting for a repair, Dad wouldn't come home until way after supper. It was like that whenever the farm needed major mechanical attention, which was pretty much always. Our farm ran on old equipment, usually discarded by other farmers.

Dad quoted his friend frequently:

"Arnold says: 'If man made it, man can fix it.'"

And "Arnold says: 'Don't force it! Get a bigger hammer."

And "Arnold says I might've amounted something if I hadn't inherited this damn farm."

The gifted mechanic didn't seem to mind Dad's company and enjoyed a lot of laughs at his expense. He continuously growled at Dad, but my father didn't seem to notice.

After straightening out more than a couple mangled car doors, Arnold joked about my father's method for "breaking in" a new car or pickup. He said: "Find two trees just barely farther apart than the width of the vehicle. Then, swing two doors wide open, one on each side, and come back between the trees."

Hey, it wasn't totally Dad's fault that we always drove standard shifts with disconnected parking brakes, and certainly not his fault that the hill in our driveway sloped directly into a mulberry tree beside the mailbox.

Arnold had a way with words. His sweet jovial phone manner belied his sometimes gruff and often profane in-person demeanor. Within his shop, there was an insiders' code of conduct, always a sense that there was more going on than a woman or child could understand.

Arnold's world felt like a private club where only men could learn the language. They didn't call the gas pedal the "accelerator." Arnold taught them to call it an "exhilarator" and the choke was a "strangler."

It seemed that all men wanted to belong to Arnold's club, while most women were happy to leave them to it. Female customers just wanted their cars to start and run well enough to pass state safety inspection.

The shop at the bottom of Ash Swamp Road contained many works in progress. Besides daily repairs on cars that people coasted to the bottom of Arnold's hill, some projects went on for years. Dad's Loraine loader transmission held down one corner of the garage for five years … at least.

One day, I asked Arnold what he was working on. Acetylene torch in hand, he said he was making stronger hooks for a trampoline used by the University of New Hampshire gymnastic team, plus, he said he was rebuilding a Packard straight-eight engine and replacing a half-axle and exhaust on a 23-foot-long Checker airport limousine. Such was the diversity of his trade. (David and I drove that same Checker to Seattle for Christmas in 1974.)

Arnold lived over his shop. He was barely 20 when he built the original stone section by himself. Then friends helped him expand it into a spacious structure of concrete blocks, locally milled lumber and tons of steel. He welded roof trusses and installed a pneumatic 'walking beam', suspended from massive Ibeams. He could hoist and move equipment to wherever he needed it within the machinist's arena.

Arnold never married. While fond of Seagram's Seven, his true passion was for heavy curvaceous cast-iron machine tools scavenged as factories upgraded their pre-war plants along the Merrimack Valley.

Dad admired Arnold's ability to fabricate parts from scratch. He said there would be no replacing his friend when he was gone. And, thus far, Dad seems to have been right. My brother Albert shared shop space with Arnold for several years and learned many of his techniques. Then he went off on his own. Albert's current repair shop isn't nearly as big as Arnold's was — as eccentric, perhaps, in the range of projects — but Arnold was more out there on the front line, dealing with the public.

Ironically, when Arnold turned 65, he finally agreed to try "going south" for the winter. Halfway there, he suffered a debilitating stroke. Sadly, he spent his final year in a nursing home not recognizing anyone, hopefully privately reminiscing about his Seagram's half-gallon under the sink and smiling at memories of telling his cronies not to worry about the filthy greasy-fingerprinted glasses they drank from.

"The alcohol sanitizes everything," he'd say.

Arnold's home and shop were in Newmarket, not far from Adams Point, where he had visited the Adams Family many times, even closer to where he helped launch the 'Cindy-Lynne' in Newfields on that fine summer's morn.

The men set up Cass's gear and, as Arnold had predicted, found it near impossible to turn the boat around on the trailer. It was too heavy, too tall, and too long to swing without pulling the flatbed out from beneath it. And, still, they'd be needing the truck and trailer to back down the ramp for the launch. After an hour of voicing opinions and repositioning the rigging, they decided it wasn't worth the effort. They'd just launch her bow first.

"It won't be the best," Cass said. "But I've seen ferro-concrete ships launched sideways, so how much trouble can we get into with this 24-footer?!" The tide was nearly high and no one wanted to miss the deep water.

Parker asked what would happen to the wooden cradle, how would they detach it from the boat, and then would it float away? Good point! They lashed it to the flatbed.

Dad agreed to stay ashore. Known to have a history of sea sickness, he said he'd gladly see them off. Besides, somebody had to save the cradle and drive the truck. He was, after all, responsible for the borrowed trailer.

"Can't be losing it down the drink now," he laughed. "Otis might forgive me, but Mary would tan my hide."

We women and children watched from the bridge as the five men got organized and let Cass call the shots. Dad sauntered toward the truck cab, doffing his cap and bowing toward us as if he were the celebrity who had made it all possible.

We applauded.

Noel hopped into the boat with Cass while Parker held the bow line.

"C'mon back!" Cass called to Dad, who began backing down the ramp. Arnold stood ready to release the winch cable and kept a keen eye on proceedings.

The plan was for Parker to hold the line and swing his boat around to the loading dock where the rest of the crew would climb aboard. Then, they would drift into the slack tide and Parker would make a little speech ending with, "I christen thee 'Cindy-Lynne'" and smash a bottle of champagne over the bow. Then, they'd fire up the engine and go off for a wee burble before tying up at Schofields' friends' private dock around the next bend in the river, about a mile away. The friends were going to serve us all a nice lunch, and it would've been really good.

But it didn't happen that way.

The launching, drifting and christening went well enough. Parker, a retired airline pilot, spoke eloquently in his loudest voice thanking everyone, especially his wife Nancy sniffling on the bridge, for their support and all before smashing the bottle with a hammer. The bubbly and glass flew appropriately all over the bow.

But when it came to the burbly motor sounds, Nope! No go!! The engine wouldn't start.

Arnold sprung into action. He tipped the cover off the engine compartment before anyone could say: "Oh Dear!" As they floated away into the channel, we lost touch with what was happening, but Arnold said something about the propeller being "shafted."

Turned out it was the propeller shaft that caused the problem. It wasn't the starter or the engine or the electrical system. They worked fine, as Arnold knew they would. But since the boat was launched bow first, its stern dipped too low and the propeller collided with a rock and bent the shaft. It was stuck and refused to turn.

There we stood watching our loved ones drift out to sea.

Dad said: "Well. I'll be!" and suggested that Nancy drive all of us over to their friends' place and see if they could rescue our crew.

So, yes. It all worked out in the end.

The friends had a motorboat and set off with a tow line. We tucked into their fried chicken and potato salad. By the time "Cindy-Lynne" was dragged into port for repairs we were already spitting watermelon seeds across the lawn.

Ahhh! The vicissitudes of the open sea.

Arnold Dennett's garage, Ash Swamp Road, Newmarket.

"It won't be the best," Cass Adams said of launching bow first.

Strawbery Banke replica gundalow "Capt. Edward Adams" moored at John Hancock Wharf, York, Maine (right beside a garden where I work). Strawbery Banke is a living-history museum in Portsmouth, N.H.

Great-
Grandmother
Louise with my
Granddaughter
Rachel

Important parts
of my life:
Uncle Eddie and
Aunt Ann Burtt
with traveling dog
'Honey'.

The way we were ...

15. FIRST DAY OF SCHOOL

When Mom dragged me downtown to start first grade I didn't have a clue.

My babysitter Janice taught me everything I knew about public school and I wasn't about to go without a fight. She had warned me about the mean women who ran the place. She said they yelled at kids all the time and had warts on their noses. I knew their names: Mrs. Quimby, Miss Odiorne, and the biggest battle-ax of all: Mrs. Beecher The First Grade Teacher.

My fingers burned and turned white at the knuckles. They would not loosen their grip on our Ford's steering wheel, not even when my mother pulled my legs straight out and I felt like I was being sucked through a porthole on Flash Gordon's space ship.

To Mom's dismay, I held on tighter and tighter and screamed "No! No! No!" My new red and green plaid dress slid up around my waist and showed my fresh yellow Sears Roebuck catalog underpants, but I didn't care. They wouldn't have me. They would NOT!

Mom coaxed and cajoled, tried to make me stop crying, begged me to behave, assured me that I HAD to go to school. She said a bell had already rung. Other mothers were standing by.

"She'll be all right," Mom told them, hoping so because she had never seen me act that way before. "She's only five," she said as an excuse. "Born in October, so she's probably starting a year earlier than she should."

Sounded good to me! I'm too young for this crap. I'll deal with it NEXT year!!

Suddenly, Mom was gone.

I flopped back across the front seat and continued crying over my woeful situation. Maybe Mom would tell them 'Sorry. She's not ready. We'd better take the extra year'. I didn't KNOW! But this was the WORST day of my pathetic little life and I just wanted to go HOME.

People walked past the car on both sides. Several looked in at me.

Another bell rang.

I just laid there.

Suddenly, there was JANICE! My babysitter.

"Hey, Janice!" She stuck her head in through the window with her hair flung out in a mess of blond curls silhouetted against the sky.

"You'd better get in there!" she hissed down onto my face. "They've gone to get the principal, Miss Odiorne. She's REALLY mean and hits kids with a ruler with a metal edge on it. If you don't do what she says, you're done for."

I sat up crying hard again and watched my babysitter run away across Main Street toward the high school. What NOW?! Should I start running, too, while I still had a chance? I could hide somewhere. In the cemetery! Right there across the street.

But here they came. Mom and an ugly tall woman who had to be Miss Odiorne. I couldn't look at her.

"You need to come into school now," the woman said firmly.

"You have to," Mom said. "I'll stay with you. Bring your lunch box."

Defeated and sick, I took Mom's hand and entered the brick building, whimpering. It was like walking into a meat locker, instantly cold and dark and smelled funny, like pee. I could feel the stink getting on me. Miss Odiorne pointed to my name 'CAROL' over a coat hook in a long narrow closet behind the 'GRADE 1' classroom.

"Put your lunch box on the floor there, under your hook," the principal said.

I tried to stop crying, but continued gasping as Mom kept her hand on my shoulder and guided me into the sunny room where Mrs. Beecher and my future classmates for the next 12 years all turned to look at me. A girl sat sobbing with her arms and head on a desk.

The fat teacher said "Good Morning!" to the principal. "This must be our missing Carol." She came forward, twisting a white handkerchief with both hands. I couldn't stand to look at her face. But what was that thing on her head? A blond hairnet?

"Please have a seat," she said.

Mom nudged me toward the one remaining empty desk and baby chair. She knelt and whispered in my ear: "You'll be okay. Try to have fun. I'll stand back there in the hall. You can see me through the door."

I tried to settle down and finally did stop crying sooner than the other sad girl, Linda Freeman, who carried on for another two days. It really helped to have my mother standing outside the door. I must have turned to look at her twenty times, just to make sure she was still there.

And then.

Who knows how long it takes for a mother to decide to duck out?

She was GONE! No Mommy in the window. So I just put my head down on the desk like Linda and cried.

Before long, I slid my chair back to breathe better with my forehead resting on the edge of the desk, still not ready to acknowledge the classroom activity around me. My eyes and nose dripped tears and snot into the wrinkles of my cotton dress. School smell wafted up from beneath my desk and made me realize it probably wasn't pee at all, but a stink they made on purpose to shine the floors.

I gazed down on my new shoes and suffered another spasm of grief. They were T-strap Mary Janes, the color of 'Ox Blood'. At least that's what it said on the can of Kiwi polish we bought to keep them nice. The left shoe was stained and dull with green stitching across the strap and along the sides, while stitching on the right shoe was white, the way it should've been.

I became acutely homesick for the farm. I ached for the ox, the blood, the manure — and especially for my baby brother Matthew, whose life I had saved two days earlier. My stomach cramped.

The previous Saturday, Mom had taken me to Exeter 'school shopping'. She did her best to cheer me up about the inevitable. She said school would be fun, meeting all the other kids and playing with them at something called 'recess'. She promised that the 'hot lunch' would be good because she had volunteered to cook the desserts.

"They're giving me all the government stuff to make something of," she said, referring to Department of Agriculture surplus food that was distributed to us unfortunate kids who had to eat 'hot lunch' and poor people in general. Mom and some other parents launched Epping's hot lunch program that year, setting up a kitchen and tables inside the nearby Grange Hall.

For several years, our own kitchen at the farm served as an auxiliary staging area for the school. You couldn't walk into our pantry because the floor and shelves were overstocked with large unlabeled tins and brown bags of desiccated everything: milk, apples, eggs, flour, sugar, rolled oats, cocoa and peanut butter. Mom spent a lot of time rattling industrial-size cake pans and cookie sheets in and out of her extra-large oven. She baked one huge dessert every morning, enough to feed 100 kids.

On our school shopping spree, Mom briefed me over hot dogs and chocolate frappes at the 'Woolworth's Five and Ten' lunch counter.

"Every day, they'll let you get your dinner box and give you a bottle of milk for morning snack. Then, at noon, you'll all walk up to the Grange for hot lunch. I'll make Whoopee Pies, apple cake and apple crisp, peanut butter cookies and chocolate cookies. I'll make both kinds of brownies — chocolate and those blond ones you like so much."

Mom tries so hard! And she had the right idea, hitting on the food and time spent outside the classroom.

At the "Five and Ten," we bought a pencil case with a sharpener hooked to the zipper and a 'Roy Rogers and Dale Evans' lunch box to hold it and the much-touted morning snack, plus some nice socks — a package of five different colors, one for each day of the week.

Then, at "Moody's," we settled on my pair of Ox-Blood 'Mary Janes'. The clerk measured my feet, commenting on their toughness and width. He guessed that I went barefoot a lot, which was true. I ALWAYS went barefoot. The shoes felt clunky.

Back home, Mom suggested that I try to "break in" my footwear by wearing them for a couple hours each day. She said I'd get blisters, but maybe they wouldn't burst if I broke them in easy enough. So during the week before school, I marched and dashed around the farm yard taming the shoes as they chafed my heels and little toes.

I was in this state of preparedness when Susie came running. "Matthew's in the manure!" she yelled. "He fell in the manure!" So I kicked my shoes into gear and ran after her toward the barn.

It was obvious what had happened. There were these two deep tractor-tire ruts where 'the men' had been backing a manure spreader in under the barn to haul manure out to the fields. Because the manure and cow pee had become soupy, the men had dumped a skim of sand to absorb the liquid, creating, instead, an unanticipated treacherous toddler trap.

Well! There was our baby brother rolling over and over in one of the deep soupy ruts as if he was a donut in the Fry-o-later. This crawler had become a green St. Patrick's Day cruller. He was totally GREEN and his manure-laden diaper was taking him down for the third time. I looked at Matthew. I looked at the sand pile that had toppled him into the rut. I looked down at my bare legs, new shoes and pretty pink socks. Mostly, it was the shoes that made me hesitate. There wasn't time to get out of them.

So I just had to do it!

"Get him out!" Susie cried.

But Matthew wasn't crying. He held his breath, eyes closed, as he rolled in the muck. I glanced one last time at my left shoe, took a breath and stepped deliberately and deeply into the cold green soup that rose immediately to my knee. I muckled onto Matthew with both hands, not sure what I was grabbing, but I wanted to make it quick, so just chucked him onto the sand at Susie's feet and stood there, one leg sunk up to the knee.

Finally, Matthew let out a wail to rival the fire siren downtown, and we ended up sharing a bathtubful of green manure water.

How I missed my little brother as I kept my head down that first day of school! I also missed 'Judy', my favorite cow, and my calf, 'Greta', and my cat 'Softpaw', and 'Tippy' the dog. I wanted to go HOME! I wanted my mother!

The teacher was trying to get the class to sing 'Fifteen Years on the Erie Canal', a song I happened to know but still wasn't willing to get involved with. Even that stupid song made me homesick!

I always sang it when I played in the cows' water trough: "I've got a mule. Her name is Sal. Fifteen years on the Erie Canal! She's a good ol' worker and a good ol' pal. FIFTEEN YEARS on the Erie Canal!!"

The trough ran the length of the tie-up and was filled from a faucet to keep the cows watered. It was a long shallow wooden box lined with metal so it wouldn't leak. I sailed scraps of board loaded with hay cut short with scissors up and down the trough under the cows' noses. Whenever they coughed on my barges or made waves by wagging their muzzles in the water, I pretended the weather had turned bad.

At some point during my first-grade ordeal, I did manage to join in on the 'Erie Canal' chorus. But I refused to use their toilets. No, I didn't go in the bushes during recess! I just held it all day for the whole year. Mrs. Beecher tried to make us all go to the toilet. They called it going 'to the basement'.

"Have you gone to the basement yet?" she would ask and then stand by as we each had to spend time in the stalls. But I just stood in there and came out when I heard other doors open.

Another thing that really bothered me about first grade was the milk they gave everyone. Two pints per student per day. One for snack. One for lunch. I was used to raw milk, not boiled milk. Everyone else drank it down, but I pretended to throw up whenever I got near the stuff. I told them it tasted "cooked." Since Mom was tight with the hot lunch people, she arranged to excuse me from milk and let me take some from home in my Roy Rogers thermos.

Life on the farm never looked better than when I was torn from it to start school. A gift to my brothers and sisters. Because I was such a First Grade twit,

The Way We Were – Part Two

Mom made sure all the other kids attended a year of kindergarten before entering public school. And, having saved Matthew's life, he was all too happy to return the favor fifteen years later when I almost rolled a tractor over onto myself.

First day of school. Susie is barefoot and delighted to have me go.

"They would NOT have me!"

Baby Matthew fell into the manure pit. I sullied new shoes.

My son Otis, same age as when I pulled his Uncle Matthew out of the drink. Here, Billy McPhee is showing him how to hammer a horseshoe.

The way we were ...

16. HARVEYS' AFFINITY FOR LAND

Following the Civil War, New England's population changed from being primarily rural and agricultural to more built-up and urban. Still, there we sat on Red Oak Hill with my great-grandfather, grandfather and Dad slowly but steadily adding acreage to the farm. First inventory I remember was 2,000 acres.

With all that acreage available, I couldn't believe I fell in love with a neighbor's property — the land we now live on — David and me, our sons, daughter-in-law and granddaughter. Before returning to Epping, we called these 30 acres of ledge-laden hillside 'Camp David'. Now 'Square One'.

It happened on Easter Sunday, 1971.

I had cried my way home after a year in Seattle where I had quit my job and boyfriend. I didn't know what the heck I would do next, but sure didn't feel like wearing a flowery hat to church and doing all the dinner and Easter stuff. So I disappeared into the woods for a morning walk, headed west with the rising sun on my back.

There were no trails and getting lost would be easy. It would suit me fine. There were worse things than ending up in West Epping.

I walked in spurts with layer upon layer of wet thawing tree detritus underfoot. I stopped frequently, listening to the forest as it crouched for its springtime pounce into summer. The woodland and I were ready to tap our resources and start over. Blue Jays squawked and critters scampered while beech trees rattled their winter-worn leaves to awaken buds that would soon release them to fall to the ground. The circular miracle of changing seasons!

I've never been afraid in New Hampshire's southern woods. Hardwoods, evergreens, meandering brooks and stonewalls make sense to me. I savor the shelter of trees, their protection from wind and scorching sun. Rain and snow falls onto my head, same as it falls onto the pines and oaks. What else could we expect in this shared environment?

It is quite straightforward and lovely. If I come to a stream I either follow it or cross it. If I encounter a swamp I try to skirt it. New England, in general, is blessed with verdant unfettered waterways. The forest around Red Oak Hill is especially enjoyable and intimate, even if you occasionally run across someone who has committed suicide.

Easter Sunday was perfect, fresh and beautiful, crackling with nature sounds — aspects of home that I had missed while living in Seattle. Foliage and black flies had yet to emerge. Patches of dirty snow and ice remained where shaded by dense juniper and fallen pine and hemlock branches. I could see quite a ways through the trees.

After a couple hours, I still wasn't anywhere. My shadow had diminished as the sun rose higher so I reckoned I was walking in circles. During frequent breaks from meandering, I ate cookies and drank water, considering which direction to go next.

The "What-the-heck-am-I-going-to-do-with-my life-now?" question hovered over me as I followed a loosely-stacked stone wall running up a mini-cliff that eventually revealed an area where timber — pine and hemlock logs — had been removed leaving just stumps. A stand of white birch swayed along the crest of a hill constrained by the uphill wall.

I immediately cherished the white-barked trees for their elegance in standing watch over the recently disturbed land. I appreciated the elevated ridge as a vantage point overlooking what appeared to be a beaver swamp in the distance a hundred feet below ledgy terraces gouged by the glacier that had formed Red Oak and Rundlett hills.

"Wow!" I exclaimed to myself. "This would be a good place to build a house."

I had no idea where I was. But it felt right.

I decided to tag the location by making a big "X" from fallen birch logs. Then I'd get Dad to take me up in the Cub and we would look down and find the spot. Then we'd know. I sat around finishing the cookies for a couple hours, viewing my love interest from various angles.

I wondered if Easter dinner was over yet and had anyone missed me? Would I ever find this place again?

So what a shock it was for me to step over the stone wall and strike out from wherever I thought I was in West Epping and — within minutes — I walked out into the familiar back Plumer Field across Nottingham Square Road and The Night Pasture! I could almost see the back of the red hay barn.

Surely, my parents owned the "Xed" birch trees! They would give me the land. I would build a house and find a new job and a new boyfriend. I'd have plenty to do. I'd be home with a purpose. No problem!

But No. My special spot didn't belong to us at all. It belonged to McPhees, who were kind enough to sell it to me.

Lessons remained to be learned about responsibilities of land ownership.

My father has always patched together and used heavy equipment — bulldozers, backhoes, loaders — that sort of thing. Not all that 'heavy' compared with what's available today. But heavy enough to clear pasture and cropland. His first digger was more of a shover: a mid-'30s John Deere Model G that he welded a bucket for and filled a cage in the back with big rocks as counter-weight. Now in his late 80s, Dad is still at it.

"I just twist my wrists this way, then that way," he says, gesturing and winking. "And eventually the trees they fall down."

Even so, Dad hasn't managed to clear it all. Bushes keep growing back along the edges of the fields and frost heaves rocks to the surface each spring. Indeed, 90 percent of our land is still undeveloped, mostly cut-over woods that my family acquired through time, cheap because it had been stripped of trees.

As you might imagine and as is typical with going modern, having societal pressure and such, there began to be 'demands' on and for our property. People were always on the doorstep asking if they could, please, buy a houselot. Developers appeared and wanted several hundred (houselots).

The tax collector forever needed more revenue and charged 12 percent interest whenever we were late, while the bank rate was 8 percent. The selectmen told Dad if he didn't like the interest rate, he should go borrow the tax money from a bank. But we already owed too much to the banks. We were always listed in the Epping Annual Report as 'delinquent'. Appraisers stopped by frequently to take second looks.

With new people moving to town, Epping schools were beyond crowded and the two buses wouldn't start on cold days. Nobody could afford to buy a new bus. And, even when they did, it wasn't much better. We still broke down and were late for class. The school board wanted money for books.

Selectmen decided to tear up Main Street and install sewer pipes. Briefly, everyone had their trash picked up at the curb. But we didn't have real curbs and the town suspended service.

Dad went into politics to straighten things out. He became a selectman while we helped fulfill his vision for 'going big' on the farm. Everyone went into debt. It was 'The American Way'. Fistfights broke out at Town Meetings that continued for days and didn't adjourn until midnight.

Epping gained a reputation for being "contentious."

Ah! That Epping woods!! With its rotting logs and layer-upon-layer of leaves underfoot—the junipers and shrubs, baby pine, maple and oak, mighty beech, shag-bark hickory, and scraggly hemlock and cedar amid cherry and birch—black, white, gray and yellow (protected by swarms of black flies, deer flies and mosquitoes, serenaded by chirping birds, chattering squirrels, croaking frogs and toads).

We didn't know he could ride a bike. Never saw him ride it again.

Mom and Erica outside Epping Community Church where our mother is a Congregational deaconess. Grandmother Eunice Peterson was inside the van (seated in an upholstered chair) to be dropped off in Iowa as Erica and Kathy McPhee continued up Al-Can Highway to Fairbanks, where they helped our Cousin Walter manage property for 'his' bank. Also, around this time, Erica returned to Canada, Alaska and Europe on a yacht as a nanny.

Susan Lois Harvey

The way we are ...

17. ADJUSTABLE AND BOX-ENDS

The Sisters' Refrain (please) ...

Phyllis and Beth are in ruts.
They always ignore our tut-tuts.
Though they are charming
Their farming's alarming.
They're hell-bent on driving us nuts.

Now The Sisters are in a tizzy.

Susie and Erica have been managing the farm's finances in recent years. They have a special relationship with our parents who appreciate that The Sisters' 'business sense' has relieved them of worry over how to pay taxes while providing them with extra money for pleasure.

Another sister, Cindy, has been donating several thousands of dollars each year to help pay the Epping tax bill. When she turned 50 two years ago, she declared herself "miserable" living with her partner Paul McPhee — (yes, the farmer McPhee who also terminated his herd in 1987). Susie and Erica (who are genuinely good women, but don't like Paul much) offered Cindy comfort and home remodeling support during the difficult break up. Understandably, Cindy joined The Sisters' farm management team and has grown mighty within the fold of our parents' inner circle.

It's a good place to be, close to siblings and loved by parents. No doubt, Phyllis and Beth wish they were so worthy. I share their yearning and probably Matthew and Albert do, too.

But there's the tizzy.

Erica e-mailed me in New Zealand about Phyllis and Beth taking tools from the mechanic shop. She wrote: "Their behavior is making me nervous. What next? How far is this going to go?" She asked whether or not we should "stand aside and be robbed?" She wanted to know: "What to do?"

Once again, in another hemisphere halfway around the world, I find myself at a loss for words to reply. But distance is not a valid excuse. I provided few responses while we were right there in Epping last summer. Seems I spent the whole time wishing I could do or say something smart and miraculously set things straight — not just the Phyllis eviction situation or the farm's finances or land deals — but, at least, step in and tell my sisters to settle down.

If raiding each other's hayfields, hijacking their cats and rabbits, referring each other for psychiatric appraisal, and chasing themselves around with brooms, and rolling each others' boyfriends down the driveway are any signs of disharmony, I'd say my sisters weren't getting along very well last summer.

Susie, Erica and Cindy ('The Sisters') were feuding with Phyllis and Beth ('The Farmerettes'). Feuding is something other families do, but not my own. Yet, what else could I call it and how am I supposed to respond to an intercontinental plea about adjustable and box-end wrenches?

Our son Amos assures me it's all right to hang back. He says: "Mom. It's not your dog to feed."

Nephews and nieces: Jared, Nathan, Sharon, Johnathan. Grammy, Grampy, Christa on Sam, Matthew, Phyllis, Ned, Cindy, Paul & Abbie, Albert (behind). Front: David, Beth, Danny, Judie (front): Otis & Amos. Some Christmas 'Up Back' on firewood log pile.

Used this Christmas picture to Photoshop 'Cornball Committee'. We worked together to throw two 'Cornballs' – public hilltop parties, no charge, a thousand people claiming to have a good time in 1980 and 1994.

We're all there: L-R Erica, Jared, Paul, Sharon, Albert, Johnathan on Sam, Phyllis, Judie, Matthew, Beth holding Christa, Analesa, me, Susie, Tom, Nathan, Mom behind David, Ned, Dad; Front: Otis, Danny, Cindy holding Abbie on Amos.

The way we were ...

18. SPUTNIK

Hard to believe the changes since my first memories of the farm, entering the 1950s, being an apple orchard with people hanging around talking about hard cider in the packhouse, then converting to a dairy operation. And Dad being forced to accept current-use taxation in 1972. Mom never did leave us to take care of strangers' children in Boston. But our entire dairy herd of 200 (milkers and young stock) was indeed destroyed as part of the U.S. Whole Herd Termination program of 1986/87.

After that, we eight kids began looking around for real jobs, no longer depending on the farm as a safe haven.

Yet the farm has remained physically unchanged, unable to keep up and unwilling to give up. Dad continually makes fields larger, but they are still basically the same pastures and cropland, the same mixed hardwood and pine woodlands, same swamps, same mud holes in the spring. The junkyard Up Back contains the same old equipment, rusting away.

Even today, entering the 21st century, the landscape around Red Oak Hill is pretty much like it was when we tethered calves to pear trees behind the house. The core 200 acres of Applehurst Farm drapes all but the northern slope of the second-highest hill in Epping. From the crest, we can still gaze across downtown where, beneath a canopy of mature oaks and maples, houses are close together. Seacoast towns of Portsmouth and Hampton are visible to the east.

So long ago, Mom took us to Hampton Beach with the neighbors, the Schofields and Welhoffers. Back then, she claimed to love the salt water. She wore an orange floral bathing suit and floated easily, bobbing us around in the frigid Atlantic, singing "Bouncy, bouncy b-a-a-aby!!" Our favorite deep-sandy beach was in front of the Farragut Hotel where the flamboyantly gay Liberace played a gleaming white piano beside the pool. We could see him from the breakwater.

Surreal memory, since we never go to Hampton Beach anymore. Mom got over it around 1960. I don't know why. Maybe we couldn't all fit into the car any more. Or maybe she had a close call at an intersection.

The fun-loving neighbors moved away. First, the Schofields, then the Welhoffers. So there was no one left to say: "Hold still, Luv!" while rubbing oil on our backs. We didn't have our own lotion or umbrella or a sporty car like theirs. Maybe Mom couldn't manage the beach without her friends Nancy and Jean. Too many people. Too much traffic. And the Farragut was torn down long ago.

Now, when we look toward the Atlantic Ocean from Red Oak Hill, along the horizon, we see new landmarks: cooling towers at Seabrook nuclear plant and cell phone antennas, and Epping Wal-Mart's bulbous white water tank, plus the low-slung arch of the Piscataqua River Bridge connecting New Hampshire with Maine, opened in 1972.

There wasn't much interstate travel before then — at least not between New Hampshire and Maine. Nobody needed to go to Maine. We didn't even need to go to Portsmouth. And, if it weren't for Dad's political interests and Mom's itchy feet, we might have stayed right there on the farm and kept it simple like the McPhees, Allens and Sulloses.

Except for 'business' trips into town or over to see Arnold for repairs, my father spent his days taking care of the cows and destroying his ancestors' orchards. He said all the trees had to go because if the cows got loose and ate apples they would get drunk and blow up. Then, even if they could still walk after the explosion, they'd go dry. No more milk.

So, Dad blasted stumps using countless sticks of dynamite stored in boxes in a metal locker in the corner of the mechanic shop. Day after day, we blew stumps. I remember taking cover with my father under our 1946 Dodge dump truck, never quite sure how far the dirt would fly...till percussive BOOM! and a puff of smoke, dirt and rocks falling all 'round.

Worst was when there was a delay, no boom, and Dad wondered how long to wait. "Could be just a slow fuse," he'd say as we waited under the tailgate.

We were familiar with consequences of farming accidents. Had heard stories about how some guy in Nottingham lost half his face because the dynamite plug had gone punky and was barely smoldering before exploding. Or how Mr. Singer from Deerfield had cut off two fingers at his mill and glued them back on with orange shellac, but only one worked because the other had too much sawdust in it. Lots of stories about the hazards of logging and sawmilling. And, of course, tractors tipping over and killing farmers. They could roll, but just as likely, something would go wrong with the clutch. It would fail to disengage the transmission and the nose of the tractor would rise and rear over directly onto the driver.

Such were the images that entertained us as we crouched under the truck watching the bludgeoned soon-to-be splintered apple tree. To his credit, Dad is patient, at least with dynamite. He waited a good long time before approaching the usually dud charge. There would be no maiming and Mom wouldn't have to raise us alone.

With tree roots loosened, Dad dragged stumps with a John Deer Model G. He said the trick was knowing where to hook the chain so a big clump of roots would let loose without breaking apart. There were piles and swaths of stumps and roots everywhere — all around the barns and along both sides of road. My grandfather's orchards had extended across from the house and up the hill. During the first ten years of my life, the dairy cows grazed among the rotting remnant of a previous economy.

I thought we should change the name of the farm from 'Applehurst' to 'Applewas' or perhaps 'Applecursed'.

Mom's mission was to get rid of the evidence.

She enjoyed being outdoors and spent hours, days, weeks, months burning the stump piles. She had no qualms about blending babies with blazes. Mom thought it was good for us. Early in life, we learned techniques for sustaining flames with kerosene and truck tires. Seemed we were always camped with a baby carriage in the pasture, watching the fires, tossing wood to feed the glowing heart of the piles. The spring-suspension, full-canopy baby carriage held two kids. Mom draped a large bug net over the rig and cinched it with elastic to assure uninterrupted naps. I can't remember ever sleeping in the carriage, since I was always too big to be a baby.

When it came to obtaining fire permits, Mom had a special arrangement with the fire warden who sat in a tower on top of nearby Pawtuckaway Mountain in Nottingham. He had the best name: Wilder Bemis, father of Hoppy, who was a couple grades behind me. The Bemises lived in West Epping and had known both of my parents from childhood.

Wilder knew exactly what Mom was up to with her ever-burning fires. We hiked to the tower to visit him and looked at the scenery a couple times each summer. He had big maps and several sets of binoculars that he let us try, assuring us that he had seen us squatting behind the sawmill.

Mom never needed a permit because Wilder never reported her.

One of her apple-tree blazes DID go out of control, though. Mom was burning stumps Up Back in the junkyard when flames latched onto a 'retired' flatbed truck. She hollered "FIRE!" and Dad came scrambling with the first frantically-filled bucket of water. It didn't take long to do damage. Dad loves his junk, so we expected him to be upset with what was left of the old truck.

But all he said was: "Everything here is half your mother's. She can burn her half if she wants to."

I remember one night in particular, when Mom walked us up onto the hill to burn stumps after supper. It was November, 1957. We were looking for "Sputnik II," a Russian satellite carrying a pitiful dog. Boston's Channel 4 told everyone where to watch for the moving star that would not twinkle. They said it would pass over every 90 minutes, but there was no set time.

The evening was cold, but tolerable because there was no wind. Mom's crackling bonfire warmed our backsides as Susie, Matthew, Phyllis and I cocked our heads backward toward the heavens. Dad had remained at the house with babies Albert and Cindy. The immense galaxy engulfed us. Within the wide-open field, the vast Milky Way spilled from east to west. I've never been more struck by the enormity of outer space than I was that night.

'Sputnik' did eventually pass silently over us. It was chilling, realizing that Earth is very small and vulnerable. My whole fifth-grade class was supposed to watch the satellite. Our teacher, Mrs. Heilbronner, said so. I knew that each of my classmates was seeing the same thing, right then, the same way with our mouths open, heads back, necks aching, each of us shivering with our families looking up at a scary sight.

I asked myself: "Will it be this way when the Russians attack? Will we all die separated from our classmates? Just die with our families, like a bunch of turkeys looking at the sky with our mouths open?"

Everyone talked about the 'Cold War' yet I was small and separated and there was nothing I could do about any of it. My whole class was afraid of the Russians. Now that these Sputniks were looking down on us, could missiles be far behind?

"Why did they put a dog in it?" I asked Mom.

"To scare us," she said.

Wasn't the end of world. But we could see it from top of Red Oak Hill as 'Sputnik' and apple tree stumps flew.

Dad spent a lot of time blowing up apple tree stumps.

The barns, circa 1813.

The way we were ...

19. UNCLE EDDIE'S BOMB SHELTER

Note: A few things in this chapter aren't just right, like wording of the box holder ad (which I got from another bomb shelter description). Uncle Eddie may not have distributed a flyer at all. But he had his own Gestetner printing machine and was perfectly capable of doing his own advertising. Some other details are third-hand through my cousins.

'Cold War' scare tactics worked well for Uncle Eddie's new bomb shelter business. His phone started ringing off the Formica once Sputnik was launched. My uncle was quick to capitalize on America's fear of Russia. He talked a lot about Communism and was against it.

Ed began building bomb shelters. And, having experienced military combat as a Marine, he appreciated foxholes more than the rest of us closet pacifists. He recognized danger and wanted to save lives. Before any of us knew what Uncle Eddie was up to, he had used his small John Deere crawler and dynamited a big hole in his front lawn.

Aunt Ann was sworn to secrecy. She wouldn't discuss the project with children.

So I asked Mom.

"What is Uncle Eddie doing?"

"Never you mind!" she said.

So I asked Dad.

"He's enlarging the cellar where the old house used to be. It's good to have a big cellar," he said. "Just being Ed, really. Don't worry about it."

All of my aunts and uncles live nearby, except for Dad's youngest sister Agnes who married a pleasant Polish man from Connecticut and raised their family down there. Uncle Eddie and Aunt Ann live closest to Applehurst Farm and have always been most present in my life. Ann is Mom's youngest sibling. They have lived in three houses, all nearby. One of these, an old Harvey house, was rented from my parents.

When quite young, I remember heading off alone to visit Aunt Ann at the 'Belle Harvey' farm. My aunt, uncle, and baby cousin Peter lived there while building the house they settled in, just over the crest of Red Oak Hill.

Mom thought I'd be OK because it wasn't far, about a mile headed away from town. So, off I went, past Plum Ladd's house where his widowed sister "Mrs. Fisher" lived, then through the Allen Farm, its house, barn and pasture.

Maple branches overspread the long narrow lane leading to my aunt and uncle's house. Grass grew down the middle of the road. It was very quiet, except for a few chirping birds and chipmunks. Bushes encroached along the sides. Rustling leaves and shifting shadows made me jumpy. Irregular stonewalls and clouds opening and closing overhead put an edge on my solo journey. I whistled off-key as I made my way, trying not to panic and run.

THEN, I encountered my uncle's closed gate with its fresh pine boards (no doubt from our sawmill) — a barrier that had never been there before — and it read, in broad-brush scarlet letters: "NO TRESPASSIN' – NO HUNTIN' – NO FISHIN' – NO NUTHIN' – SO KEEP THE HELL OUT!"

I was initially shocked just to see the gate and even more so by the time I had analyzed the message. The word "HELL" flipped a switch to drain blood from my face. I could almost see my aunt's house through the trees, but the gate repulsed me. I could neither touch nor duck past it to enter forbidden territory.

So close! Yet, so unreachable. I panicked and raced back to my mother, breathless and weak.

Of all my conservative relatives, and there are many, Uncle Eddie is the most faithfully resistant to change to the point of causing anxiety. At least that's the way he seems to me. He sure knew how to alarm THIS kid! He didn't mean to. He just did and continued to serve this stabilizing purpose through my life – like a keel as I went off to college amid cautions against turning into a Communist – "Stay on the straight and narrow. Don't get brainwashed."

And, with the launch of his "Anti-Blast" business, Uncle Eddie began doing his best to scare adults, too. He placed ads in 'The Exeter News-Letter' and covered area towns with a box-holder flyer. His carefully drafted logo showed a family standing under a 'FALLOUT' umbrella. Bold typewritten text quoted Soviet Premier Nikita Khruschev's "We will bury you!" and went on to invite the public to visit the model bomb shelter attached to my aunt, uncle and four cousins' basement.

Going public like that must've been hard for Uncle Eddie. I'm sure his worst nightmare involved living his final hours underground, squeezed between relatives who hadn't bothered to build their own damn shelters.

My mother can sound very disgusted when she is. In her most don't-bother-yourself voice, she told her sister Ann not to worry about it.

"You couldn't PAY me to stay in that thing!" Mom said. "People aren't meant to be underground until they're already dead." Her comment eased tension after my mother caught Ann's nervous glance toward what turned out to be a shotgun behind the kitchen door. Ann's eyes betrayed Ed's turmoil in wondering if he'd be able to shoot his in-laws if he had to.

"There just isn't room enough for all of us!" Ann said. "I'm glad you understand."

Ed's "Anti-Blast" flyer promised to provide safe living space for a family, strong enough to withstand an atomic blast with its double-reinforced block walls, storage for a year's worth of water, dry-packaged rations, and battery-operated air exchange and filter. Fallout protection suits and first aid kits were optional, as was a flush toilet and telephone (which were not guaranteed to work), but would complete the "Duck and Cover" package for just $4,999, "discreetly installed."

Sad to say, we'll never know whether or not Uncle Eddie ever sold a bomb shelter. We don't even know if his old shelter still exists. He was that discreet, never revealing much information and (contrary to what he probably thought) we didn't keep a very close eye on him. But the rumors certainly could've been true. People were apt to believe most anything during that Cold War era. Citizens were expected to watch the sky and report anything "out of the ordinary."

We heard that Ed was building a bomb shelter for William Loeb, the notorious right-wing publisher of New Hampshire's largest newspaper, 'The Manchester Union-Leader'. Cousin Peter let it slip that his dad had been spending a lot of time in Pride's Crossing, Massachusetts (the out-of-state publisher's hometown).

"But it's a secret." Hush! Hush!

Funny thing about that gate that I didn't dare open so many years ago:

Now in 2007, my sister Cindy was in court last month testifying about that same barrier. The question was whether or not a police officer should have passed through the gate while investigating a murder.

Books will be written about this one! Our dominatrix neighbor, Sheila LaBarre, is in prison for (allegedly) killing one of her boyfriends and burning his body on the lawn of that same house that Aunt Ann and Uncle Eddie lived in. If it is determined that the policeman entered without a warrant, much of the evidence could be thrown out on a technicality.

Doesn't seem right, does it? Two of her other boyfriends are still missing and somebody's toes were found in the septic tank.

A month of helicopter infrared scans and canine searches can't prove the men are dead, but neither can they be proven alive. Shouldn't one unquestionably dead person be enough to prevent Sheila from ever dropping by again for sweet corn?

Everyone knows poor Kenny's a goner. They even have the young man practically bidding farewell on the new Epping Wal-Mart's surveillance tapes. Day before Kenny's disappearance, Sheila was filmed pushing the sullen bruised victim in a wheelchair, through Sporting Goods, with two cans of camp stove fuel on his lap.

Wouldn't it be strange if the gate that was built in 1957 to repel Russians and nieces is used 50 years later to let a murderess escape?

Page B8 • NEW HAMPSHIRE UNION LEADER • Wednesday, May 13, 2009 State / Entertainn

Former owner may bid on LaBarre's horror hous

◆**Auction May 28:** Dan Harvey says he would like to reunite the property with the rest of the family's farm.

By JASON SCHREIBER
Union Leader Correspondent

EPPING — Some might call it a house of horrors, but that doesn't faze Dan Harvey.

The 88-year-old lifelong Epping farmer used to own the now infamous farmhouse where Sheila LaBarre savagely murdered two men and then burned the dismembered body of at least one of them.

When the foreclosed property at 70 Oak Hill Lane goes up for auction May 28, Harvey's family plans to make a bid. Harvey said he'd like to see the LaBarre farm returned to the more than 2,000 acres of land that's been in the Harvey family for generations.

The circa 1777 Cape-style farm.

LaBarre admitted to killing Kenneth Countie, 24, and Michael Deloge, 38, but claimed she was insane at the time. A jury disagreed, finding her sane and convicting her of first-degree murder. She's now serving two life sentences in a Massachusetts prison.

Harvey believes the farmhouse where the walls were once splattered with blood should be saved and not torn down.

"I don't see anything you can't fix up. It's a good place for a hired-man shack," he said.

Harvey owned the farmhouse and its nearly 120 acres before selling the property in 1962 for $14,000 to the now deceased Dr. Wilfred LaBarre, the man whom

JASON SCHREIBER PHOTO

Lifelong Epping farmer Dan Harvey, 88, used to own the now infamous farmhouse where Sheila LaBarre savagely murdered two men.

at $860,900, Dan Harvey said he'd like to see the house and barn remain and the property

sells at her farm.

Like her father, she said she'd like to see the house restored

house and several vehicles with smashed windows parked on the property.

If she were looking to buy property and the price were right, Brenda Moore said she'd consider purchasing it.

"People buy homes all the time that people have passed away in. I don't think anybody would be bothered by that," said Moore, who lives on Red Oak Hill, not far from the LaBarre property.

Not everyone is so sure the property will be an easy sell.

When he heard last week that the property would be auctioned off, local historian Matthew Thomas said he couldn't help but wonder, "Who in their right mind would want to live there."

"I certainly wouldn't want to buy it knowing what happened

Barre murders one of th
heinous crimes in Rock
County, described Eppi
"cursed town" plagued
edies throughout its hist

While the gruesome
were highly publicized
now the basis of a tru
book called "Wicked Int
The Sheila LaBarre Mur
author Kevin Flynn,
said he believes the
case could one day be an
forgotten chapter in E
history.

Residents move in an
towns so quickly these d
it's hard for most to und
the local history.

"People really just
these towns. They're so
nected and history is not
good or bad," he said
murder case) will be in
nals of Epping history, b

Article written by Cousin Jason Schreiber, son of Eric and Carol, all of Fremont next to Brookvale Farm.

Deceased neighbors Doctor Wilfred LaBarre (top)
Plum Ladd (left) always tooted when he passed our house.

Below: Matthew trying to start the 'Hay Crew' truck.

The way we were ...

20. N.H. DAIRY INDUSTRY -- MANAGEMENT

Our Applehurst Farm is not typical of the American dairy experience — certainly way different from America's western agribusiness. I'm pretty sure we are unique even among other dairy farms in New Hampshire. Just around us in Epping, the McPhees, Sulloses, Allens and we had different management styles, despite our belonging to the same 4-H club and selling to the same milk company. My Peterson grandparents' Brookvale Farm had its distinct character, as well.

Historically, individuality is a Yankee birthright.

Most New England families got rid of their cows and took off-farm jobs after the Civil War. Entering the 20th century, further advances in technology led to radical redefinition of 'family farms'. Families on dairy farms were forced to choose between getting big or getting out of the business. Rather than spend money to comply with government and milk company expectations, many farmers gave up and joined the mainstream workforce.

Understandably, dairymen across the United States shied away from daunting long-term investments lacking solid repayment plans.

The intimidating list of "essentials" included upgrades of: buildings and cow yards, tractors, modern crop and other equipment, pipelines, artificial insemination programs and high-production infrastructure for refrigeration, feed and waste management, forage crops, and hybrid corn, fertilizers and herbicides. Most farmers quit well before loose-housing and bulk milking systems, and other modernization.

Perennially low milk prices did not warrant the debt. In New Hampshire, the number of dairy farms dropped from 10,572 in 1940 to the current 125, according to 2008 government data, which also claims that, across the nation, 24 dairies continue to close each day. I believe six more New Hampshire dairy farms sold out this summer.

The courageous few who carry on still do so with an independent flair and disdain for conformity. New Hampshire dairy farmers do NOT share blueprints to build their barns and milking sheds, as they do in New Zealand where management and facilities are very similar from farm to farm. Yankee farmers pride themselves on using their own wits to sustain their families within a complex socially-, economically-, and environmentally-fragile world.

That's the 'lifestyle' aspect of American agriculture, the part that has no price tag and is mourned as small family farms are broken and pushed to extinction.

The Cooperative Extension Service suggests that, among the handful of dairy farms remaining in New Hampshire, the 'average' herd size is 115 milkers (ranging from herds with 15 cows to one farm with 1,000). To me, the skimpy data points to DIVERSITY, rather than anything typical. I don't believe there is an 'average' New Hampshire farm.

That said, I suggest that we Harveys DID indeed have much in common with the other 14,000 American farm families who left the market in 1987. At that time, the U.S. Department of Agriculture profiled and tagged farms that were similarly expendable. They trained and dispatched staff to counsel a hundred of us in New Hampshire, convincing 59 families to stop milking cows. We, the terminated losers, would not be missed by urban milk drinkers as long as supermarket refrigerators remain stocked.

Unlike New Zealand farmers who .....

All right, already!

Enough about New Zealand!!

My sister-in-law says: "Nobody wants to hear about how much better things are in another country. Just stick to our own farm."

So. Okay.

Although I decided long ago NOT to stick to our own farm, here is an abbreviated list of Kiwi dairy advantages:

— New Zealand dairy farmers appear to readily adopt industry standards, practices and plant configurations. They collaborate on best practices for efficiency, following recommendations from cooperative milk companies with specified expectations.

Contrast to: Northeastern Yankee dairymen who shun conformity, preferring to drape their banners of Independence over everything right down to the milk bucket. Good old-fashioned New England dairymen don't want to run their farms the same as their neighbors. Compliance is bad; (obstinate) individuality is good.

— New Zealand controls 25 percent of the world dairy export market. With just 4.2 million people and 5.5 million cows, New Zealand sells 95 percent of its milk overseas, mostly as powder. It is by far the world leader in per capita dairy export (because there are more cows than people). This year, the world's largest milk powder dryer was opened near us on the South Island, and until recently, the North Island laid claim to the world's largest cheese factory – all for export dairy products.

Contrast to: United States exports less than 10 percent of its dairy products. Most milk is used and consumed within the country. This is good for Americans because food prices are minimally impacted by export demand. Food in the U.S. is much cheaper than in New Zealand, where you might expect otherwise because agriculture is such a strong segment of the economy. It's a common complaint by locals. All the good stuff is exported and what's left brings top dollar to match what farmers get for export. Kiwi processors cut the citizens no slack. U.S. farmers lack that leverage and receive low prices for milk.

Update, 4/7/2010 from 'The Washington Post' article about small Ohio producer: "In the first three months of 2009, conventional dairy farmers saw prices plummet by 30 percent, due in large part to falling exports …"

— New Zealand dairy farmers seem to mirror each others' responses to ever-changing situations

like drought or downpour; mastitis or lameness; crop conditions; fluctuating milk prices; too much of this, too little of that; and dozens of other factors that I know nothing about. As a relief milker, I have found it easy to roam between farms because facilities and procedures are so standardized.

— Most New Zealand dairy farmers depend on specialized contractors to do what American farmers generally do themselves. For example, Kiwi contractors own fleets of trucks and custom-made equipment. They are hired to expertly plant and harvest crops or erect and maintain fences, troughs, drainage and irrigation systems. Sometimes, 'contract' and 'share' milkers assume responsibility for herds. There are several arrangements unique to New Zealand, such as '50-50' share milking.

Contrast to: Yankee farms managed by individual families, not particularly collaborative, holding cards tight to chest. We have the U.S. Department of Agriculture to guide us. Who needs more?

Update, 4/7/2010 from 'The Washington Post' article about small Ohio producer: "(U.S.) per-capita consumption has plummeted 30 percent since 1970, according to government figures. Many American consumers long ago accepted milk as a bland and watery, if nutritious, obligation. But Taylor's evangelism is aimed at not only consumers but also his fellow milk producers. 'The big lie is that all milk is the same, and therefore everyone gets paid the same for producing it,' he said, referring to the government guidelines that regulate the price most farmers receive for their milk. 'That is what is crippling the whole. It keeps the entrepreneurs and the little guys from rising up."

— Nearly all Kiwi dairy cows are artificially inseminated during the months of October and November (late spring) to calve simultaneously the following spring. So (nearly) all herds are dry during the winter. This gives New Zealand farmers an approximate three-month break from milking.

Contrast to: I've never heard of this sort of breeding program in the United States. Seems American dairy herds are bred to produce milk year-round, rather than have all the cows calve at the same time. (But I certainly don't know it all!) But, yesterday, at a New Zealand Bluegrass festival, I met a local dairy farmer who had traveled and worked in China and the United States researching the dairy industry. He operates a 650-milker farm in Southland. He said there was a Connecticut farmer (milking 25 cows, I think) who wanted help with rotational grazing. But, said the Southlander: "He wasn't doing it right and wouldn't listen … wasn't willing to change." He said Quebec and New England dairy farms seemed "backward." (He's not the only N.Z. dairy farmer who has told me this.)

— In New Zealand, many sheep stations have been converted for dairy in recent years. I have worked with herds of 300-700 cows on seven different Southland dairy farms. Each installation is an engineered masterpiece using the latest technology and architecture — unique, yet basically herringbone, rotary or robotic in design — all leading to towering stainless steel cooler vats that any milk tanker truck driver can approach, suck dry, and exit within just a few minutes.

Contrast to: Our New Hampshire farm, where tank truck had to reverse down Red Oak Hill Road (one-eighth-mile) into our driveway where it sometimes hit the mulberry tree.

— New Zealand's industry-changing inventions include the 'herringbone' milking parlor (by Ron Sharp in 1953) and the 'rotary' shed (by Merv Hicks in 1967), plus movable electric fencing, rotational grazing, export refrigeration, 'calf-a-teria'-style feeding, and other advanced dairy practices. New Zealand has done its part to promote high-quality milk production throughout Europe, North and South America, India and China.

I have limited experience on dairy farms. But, from what I've seen, it seems to that a common denominator is the farmers' genuine love for cows.

Phyllis certainly loves her cows and any herd owner I've met in New Zealand shares her fondness for their lactating beasts. They take good care of their herds and the cows take good care of the owners. More milk in the bucket; more money in the dairymen's pockets – more so in Southland (NZ) than New England (USA).

I see my family's farm withering back home while New Zealand dairies can't build herds and sheds fast enough to meet market demand. Kiwi cooperatives are not government regulated. They independently monitor their suppliers to meet production quotas and quality standards. It seems to me that New Zealand has grown its dairy industry gracefully with healthy cows and wealthy farmers providing a quality product.

Contrast that with American dairy farmers who struggle to meet industry demands. They seem to work harder and enjoy it less because they aren't paid enough for their product and their herds are similarly stressed. From what I've experienced, I must say it is easier (and more profitable) to milk 400-700 New Zealand cows than it ever was to milk our 100 in Epping.

According to New Hampshire statistics provided by the Extension Service, 2007:

"With total (average) farm receipts at $363,350 and farm expenses at $336,530, annual net income averages $26,820 per (New Hampshire) farm" ... which is expected to pay the owner's and unpaid family labor and other management costs.

That means a farm family (at least two people) is supposed to manage a 115-cow dairy operation 365 days a year for $27,000. The Extension Service calculates that $60,000 to $77,000 would be more reasonable compensation for today's 'unpaid labor and management'.

The government agency goes on:

"The obvious conclusion is that this situation IS NOT SUSTAINABLE. If it exists for many years, farms start to show signs of deterioration and farm families become stressed as they are unable to make sufficient income to cover living expenses. In the long-run, dairy farmers exit the industry and seek employment elsewhere. The land and farm property are sold and often become used for other ventures."

As I write, my sisters Susie, Erica and Cindy have hired carpenters to make sure cows NEVER return to

Applehurst Farm. They are completing long-overdue repairs and improvements to the barns. One notion is to convert the buildings for boat storage.

The 60-years-worth of concrete — the cow yard that was enclosed in 1970 and the contiguous stanchion barn of the 1960s and original tie-up hay barn, dated 1813, with three upgraded milk rooms, evolving from jugs cooled ice water to bulk handling – now, all tagged for non-farm use.

Just a sign of hard-hearted economic times, I guess. A dry-dock marina on top of Red Oak Hill. Not for our boats. Other people's pay-to-store boats. Similar to The Sisters' recent transformation of the home we grew up into an apartment house.

Not for our family.

Other people's families.

Phyllis's cows in junkyard section of Night Pasture, 2008.

Water under the bridge: Cousins Danny Harvey and Amos Clapp passing under Main Street Town Hall Bridge on way to winning 2010 Lamprey River Canoe Race.

The way we were ...

21. HE SHOULD'VE USED THE POND

The dream begins and ends the same way. I can't breathe and Mommy is crying. Sometimes, my mother is actually there, sitting on the edge of my bed gently shaking me awake. She understands why I'm having this recurring nightmare as an eight-year-old. She tells me everything is okay. She is not crying.

But in the dream, Mommy is the scared one and she IS crying. She runs away and I follow. It starts out deep inside the old barn, very dark, in beyond the cows, where Mommy and I have pitched fresh hay into the manger. Airborne dust makes the morning sunlight appear solid as we walk through it past the windows, headed for the door. I anticipate blinding light, fresh air. I'm hungry for television — Saturday morning cartoons, 'Fury', 'My Friend Flicka', 'Little Rascals' — the best part of each week.

We're almost out of the barn.

Oh NO! Mommy shoves me back.

"Whaaaaa!" she shrieks. What's happening? I'm startled. Russell our hired man, definitely Russell, his back to us, bent over the milk tank. Cats yowling, terrified yowling that makes my chest tighten. I begin gasping from the stink of tomcat, the desperate keening, the stench of cat mess, Mommy's fear. Two burlap bags seem alive on the wet milkroom floor. Something rolls and cries within the bulging rough brown fabric. Russell has an aluminum sawdust shovel.

I don't understand.

Suddenly, Mommy growls "Arrrrghhh!!" and runs past Russell out the door. I lurch forward, too, and trail after her in a race for the house. I can't catch my breath. It's the way I cry. It hurts. My throat is so tight all I do is spasmodically inhale, unable to let anything out. Mommy is crying. We're in the bathroom.

It's awful.

This is where I usually wake up.

In real life, when it actually happened, Susie, Matthew and toddler Phyllis huddled outside the bathroom watching us through the door. Before long, Susie was clinging to Mommy, too, sobbing.

What my mother and I saw in the barn was the hired man drowning cats in one of the water-filled milk tanks. He had caught four kittens, their mother and the old yellow tom, stuffed them, clawing, into sacks and was submerging them under a heavy wooden grate inside the ice-rimmed tank. He used the shovel to hold the grate down until the cats were dead. No doubt, he wished he could kill them all at the same time, but the rack wasn't wide enough. Certainly, he wished my mother hadn't come along between batches.

Russell was murdering our cats and our mother didn't stop him. What did it mean? Was it too late to save Buttercup and Softpaw? We kids were confused, but clear about one thing: NO! We couldn't go out there. It was the first time we saw Mom so paralyzed. What if the hired man caught us and threw us into the tank, too?

Dad was off somewhere, but far from uninvolved, since he was the one who told Russell to get rid of the cats.

"Daddy said to do it," explained Mom, regaining her composure. "But he should've used the POND!"

That was back before tanker trucks. Twice a week, Johnny The Milkman backed up to the milk room with his refrigerated box truck and yanked a dozen ten-gallon jugs from our two tanks of ice water. The cooler Russell used for the cats contained water but no milk jugs.

It could've been worse.

Still, it was the thought that counted. That and the fact that we buried six cats behind the rhubarb.

Things weren't right for a long time after that. Eventually, I got over my nightmares. Still, it felt like Susie and I had been shoved into growing up sooner than we intended. I'm the oldest, two years ahead of her. Still, my assertive blonde sister seldom acknowledged my seniority and always seemed bigger.

The only time I accepted Susie as physically smaller was when our babysitter, June Allen, was married at Epping Community Church. She chose Susie instead of me to be her flower girl.

"You're too big," June said, not meaning to hurt my feelings. Susie didn't know a 'pew' from a 'ring bearer', but I had to admit she was a cute little carnation carrier.

Sibling rivalry aside, Susie and I were each other's closest companions and Mom expected both of us to help out. Besides, we had our chores. Susie and I needed to work as a team to finish the work. 'Chores' included twice-daily feeding and cleaning up after the calves, chickens and pigs, and helping with the cows. We threw bales of hay down from the loft and spread it in front of the milkers, shoveled grain and sawdust, hoed manure into the gutters, wheelbarrowed silage, lugged and poured buckets of warm milk into a large stainless strainer to be cooled.

Eventually, as the younger kids grew, Susie and I exercised our privilege of ordering them around, yelling at Matthew and Phyllis to hurry with the hay or we'd miss the school bus. Then Albert and Cindy grew tall enough to lift buckets off the floor and we could boss them around, too.

Six years later, our workforce was expanded by 'The Babies' Beth and Erica. By then, I was in high school, not overly burdened by my brothers and sisters.

What I'm trying to explain is my birth-order pairing with Susie. As the two oldest sisters, we always had to work together and shared the growing-up process into our twenties. And, though we've never been best friends, even now we continue comparing notes, more or less out of necessity.

"Why did you cry?" I asked Susie, back then, after the cat-drowning incident. "You didn't even know what happened and you cried."

"I couldn't help it," she said so many years ago.

And, of course, I couldn't help it either, just seeing our mother in distress. We struggled to make sense of contradictions that became apparent following the calamity. First off, our special cats 'Buttercup' and 'Softpaw' had been spared because Mom had pointed them out as the two good cats. And, yes, it had been our own tearful traumatized mother who had told Dad to get rid of the kittens and whatever other cats he could catch.

So, the question Susie and I never really answered is: WHY was Mom so distraught?

We kids soon became used to eating stew-beef and hamburger identified by name, the result of having our pets murdered. But our mother was MUCH older than us and had lived with agricultural death and disaster all her life. She was long past being upset by the offing of a few barn animals.

Probably she was simply startled by the milkroom scene and ran. Once, I startled myself by seeing Dad pee in the gutter and ran away, just like Mom did. But it didn't make me cry.

More likely, she was upset because she could have killed the cats better herself.

My mother used the threat of doing things herself as a disciplinary tool. We grew up listening to Mom's muted moan about having to do everything herself.

"Why do I have to do everything myself?" she'd say to no one in particular. And then, becoming more specific, she would look directly at one of us and snap: "Don't bother! I'll do it MYSELF!" Then she'd get in the way, blocking us sluggards from doing whatever she had requested.

It made me mad always having to respond quickly to her passive orders. It made me feel especially stupid pleading for a second chance to be good. I knew she got it from Grammy. One of Petersons' hired men told me: "Your grandmother is a hard woman. She'd always say: 'I'll go get started on that myself' when all we wanted to do was digest our dinner for a bit."

Susie and I were too young to figure out the cat thing. We lacked the vocabulary and knew nothing about Mom's being pregnant or over tired.

Alas, I'll never forget the desperation in our mother's voice as she cried: "He should've used the POND!"

Mom oversees yet another generation. My boys (Otis & Amos) being taught how to run a motorcycle by Cindy (Cynthia Ellen) and Mom. Now, that little kid on the back of the bike has married and delivered a great-granddaughter.

Epping has been good to me. I had privilege of leading an alumni effort to revitalize our old high school through 'The Watson Academy Project'. Volunteers reconstructed the bell tower, had a parade, hosted the governor twice. Shown here weathervane donated by David and me.

The way we were ...

22. DAD IN BOSTON HARBOR

"Mommy's on the phone with Eddie Lavoie," Susie told me as I tried to make a newborn calf drink milk from a bucket.

"So what!" I responded as the calf frantically sucked on all four fingers of my right hand while I lowered its nose into the warm milk.

"So, he's the POLICE!" my sister said. "And it's about Daddy. She's trying to get Schofields' number at their new place."

The little heifer lurched forward, as they often do, and butted the bucket to get more milk as if it was nudging its mother's udder. It got more milk all right! Got it all over ME, straight up into my face.

I looked at Susie with milk dripping from my eyebrows and understood her concern. Dad had flown the blue plane down to Massachusetts that morning and hadn't come back. Mom already asked her sister, our Aunt Ann, to drive to Newmarket and find David "Dizzy" Dean to milk the cows. So Ann did and then stood in the walkway behind the dairy tie-up, anxiously waiting until Dizzy dumped enough milk into a bucket for her to lug to the milk room.

"At a time like this, it's good to have something to do," she told the hired man who worked for us three days a week. We just didn't know which three. Mom said Dizzy wasn't dependable.

Now it was getting really late, almost dark, George Washington's Birthday, February 22, and very cold. There were about four inches of crusty snow on the ground, icy from a month of midday thawing, then refreezing at night.

Wherever he was, Dad was on skis. Or at least the Aeronca "Defender" was. The four-foot-long aluminum skis made the 1940 airplane look young, somehow more jaunty, than when it ran on summertime wheels. The tandem two-seater had a handsome paint job — multiple coats of "dope" over fabric — royal blue fuselage, orange wings and ailerons, flashy lightning bolts the whole length of both sides, prop to tail.

But where was this colorful sight now? Our "Defender" on its shiny javelin skis? And where was Dad? I was only ten. Young, but not so young I didn't know what was going on. I knew why Ann and Dizzy had been called. And I knew for sure that if it got any darker, Dad couldn't find his way home. He wouldn't be able to judge distances and he might crash, again.

"She thinks Eddie Lavoie has Schofields' number?" I asked, sharing the situation with my closest sister. We guessed that Mom called Epping's chief and one-man police force in case there had to be a search.

"Searches" were fairly common back then and seemed to involve the whole community. Anyone could join a search and everyone was urged to do so. Dad took the plane up regularly to look for lost things like missing children or wandering old folks, but usually it was to search for escaped cows, either ours or McPhees'. Last year, Otis Goodrich came up the hill and asked Dad if he'd take him to search for two of my classmates, Robert and Jerry, who didn't come home by sunset.

"Sure!" Dad said. "But have you noticed how dark it is?"

Otis smiled, whipped out his new six-D-battery flashlight, and said "Let's go!" It was a silly exercise, but gave the town something to talk about for a week. They actually FOUND Robert and Jerry from the air when the boys flashed "S-O-S" back at them from where they had decided to camp beside the Lamprey River.

Regrettably, not all community-wide searches turned out so well. The worst was when the minister's seven-year-old son was found IN the Lamprey a week after he fell through the ice. And, of course, there were the widely-publicized ill-fated searches for kidnapped girls and suicidal boys. A ten-year-old girl from nearby Exeter went missing while walking home from school. Newspapers, radio stations and TV news sounded the alarm and asked citizens to look for little Patti. But she was never found, though they thought they were onto something when a hot tip led to an Epping gravel pit.

Another search ended tragically when everyone looked for Kenny Plaistridge (name changed) and, after four anguished days, somebody looked up to find him hanging in a tree. A boy two years behind me in high school accidentally killed himself while hunting. Can't remember his name. He was a new kid from Massachusetts.

There was also this hunt in 1960 for an 18-year-old from Manchester, a girl named "Sandra Vallard." Her name haunted me every night as I dragged 15 bales of hay from the black recesses of the mow and toppled them down through a hole in the floor to the cows below. I was very scared to be up there alone. Hairs on the back of my neck did, indeed, stand up every time the barn creaked. I imagined ways to use my sharp hay hook as a weapon. The television declared that whoever stole poor Sandra could be hiding "anywhere within a 25-mile radius of Manchester." Epping was within that range and the hay mow could be "anywhere." It felt like the guy was hiding, watching me, without anyone realizing how very scared I was.

Anyway, back then, "searches" weren't left to the experts. A sense of responsibility, everyone looking out for one another, extended even to us in the rural areas. Epping firefighters and rescue people were volunteers, neighbors. So, it wasn't a big deal for Mom to call for help at such an early stage of her concern for Dad.

Susie and I left the calf fighting with its bucket and went into the house to find our mother off the phone.

"Eddie Lavoie said he'll see what he can find out and call back," Mom said. She sat at the kitchen table with our black rotary-dial telephone on her lap. "He'll get Parker's number from the phone company, then call and see if they know anything. Daddy should've been back before now." Mom was worried and sent us outside to listen for the Defender.

Susie and I stood for awhile on the front lawn and wandered out onto the road, then into the field, studying the southern sky beyond downtown Epping. Usually, the rosy rays of sunset above the barn were a welcome sign of day's end. But not that night. Cold late-afternoon shadows rushed over us with unnerving speed, spreading a blanket of frigid darkness. There was no distant drone of incoming aircraft. No return call from the chief of police either. Nothing we could do, really.

Not until Eddie Lavoie's dark blue Buick with "603" on the license plate pulled into the driveway.

Susie and I ran in from the pasture. Mom came from the house with our crying-baby sister Phyllis on her hip. The tall chief climbed out of his car wearing his regular clothes — work pants and plaid winter jacket. His real job was running a propane delivery business. Our chief saved his one uniform for court, funerals, and to guard the moderator during Town Meeting.

"Sorry it took so long, Louise," the chief began, explaining that he got through to Parker Schofield okay and, yes, Dad had been there in Scituate and left at about four. "But, then, I left it with Parker. He said he knows the Coast Guard and he'd call them. He said he has your number and will call here."

As if on cue, the phone began ringing and we all rushed for the kitchen.

"Hello!" Mom answered. "Oh Parker! It's Parker." Her face fell, reflecting what she was hearing over the line from Massachusetts. When she spoke next her voice had changed to a somber tone. "Yes. Yes. Oh, that would be good. Yes, please, and John, too … Yes. He can milk in the morning. That would be good … Okay … Yep. Yep … All right. He's here. Thanks, Parker … See you."

She handed the phone to Chief Lavoie and told us Dad was OK and we should get ready for bed. Dad would be home in the morning, she said.

But that's not what she'd been told. What Parker Schofield told our mother that night was that the Coast Guard station in Hull was responding to reports of airplane wreckage on Outer Brewster Island at the entrance of Boston Harbor. They were sending a boat to investigate. Parker had also insisted on driving up to the farm, leaving right then, and he was bringing Uncle John with him. It would take at least three hours for them to arrive.

For several months, Uncle John and the young Floridian, Noel Gosnell, who forgot he was supposed to go home for the winter, had been living in a motel in Scituate while helping build the Schofields' new house. Parker and John arrived in Epping at one in the morning, and for some odd reason, Mom served them "the worst-tasting coffee" she'd ever made: "Chock Full of Nuts."

"All I could think of was that awful coffee!" Mom recalled, plus she said there had been so much going on that she forgot about the furnace, another sure sign that she was REALLY upset.

"Dan always tends the fire and, by the time I remembered and threw more chunks in there, it was too late. It just went out. It was such a cold night with the wind blowing, the house cooled right down." Dad was the only person who could keep the furnace burning, since it was his own (finicky) design. He had fabricated it himself. The firebox door displayed in steel bead: "Harvey's Inferno."

Mom was more relieved than she meant to be when the men pulled into our driveway during the wee hours. She said she couldn't help crying when they came through the door. We, the kids, were sleeping, buried under all the good blankets.

Mom said Parker and John told her to go to bed, too, but the Chock Full of Nuts kept her awake — that and wondering if they'd ever get the furnace going — and hoping she had scrounged enough spare blankets to keep her friends warm — and wondering if Dad was dead or alive. Sometime, toward morning, Mom drifted off.

In the meantime, Daniel Webster Harvey was making memories. He lived to tell the story many times:

"So, I was paddling my way along toward home when, at 3,000 feet, the engine threw up and stopped just this side of Boston Harbor." He said he "meant" to fill up with gas at Beverly Airport, but didn't get that far.

"Cold weather uses more gas than you realize," he said, with a further disclaimer that the floating gauge had frozen stuck showing a higher than true level of fuel. He could have topped off at Hyannis Airport where he had delivered a side of bacon to his older brother, Edward.

"But it was bare ground around the pumps and I didn't want to scuff up the skis." So he headed north, "like an Alaskan bush pilot" delivering Uncle John's mail and a second slab of bacon to the Schofields in Scituate. Dad landed in a frozen marsh beside Parker and Nancy's house. They visited awhile and then he took off.

"I would've filled up in Beverly and just about made it home before dark," he said. But he ran out of gas instead.

"I said: 'Uh-oh!' What WAS that last island I just passed?' I swung her 'round, just coasting, nuthin' happening with the engine, and landed her kinda normal-like on a high flat bluff on top of the island, 50 feet or so above the sea."

Everything seemed to be okay, him and the plane. So he got out and looked around "like Robinson Crusoe" and figured he'd better build a fire as a signal and also because he was getting cold. "By accident" he had a book of matches in his denim winter jacket pocket.

"There were lots of boats going by, headed into the harbor. But they couldn't tell by looking at me that I was stranded because the airplane was around the other side."

Dad said he wandered stomping his feet and swinging his arms to warm up. He discovered concrete World War II army barracks with leftover coal that he piled up with aspen branches. Then, when he went to soak a spare glove with gas from the plane's sediment bowl "to make sure (he) got a good conflagration going," he realized that he had laid up his fire where it couldn't be seen from shore.

"So, I had to gather it all up and set 'er again 'round the other side of the barracks." He said he kept thinking of himself as John Dame, a man who slept in his boat and drifted when heavy fog settled on him in nearby Portsmouth Harbor, just up the coast. "Dame just said: 'Ho! Hum! I guess I'll have to spend the night,' and curled up in the bow of the boat."

Dad prepared to sit tight until morning. “I had Saltine crackers in my pocket, so ate them and propped myself up on a broken settee there and relaxed by the fire.”

At about three o’clock, he was awakened by men hollering at him from a boat. It was the Coast Guard with flashlights in a small landing craft.

“I stirred myself and sallied forth to greet my rescuers.” They talked through an electronic megaphone. “STAY WHERE YOU ARE!” they ordered. “But I didn’t. I just kept walking toward where they were coming in. ‘WHERE ARE THE OTHERS?’” they wanted to know because they were ready to deal with a plane crash.

In the end, Dad was saved and taken back to the Hull Coast Guard Station where he waited until six to call home. Word spread fast around Epping and “all across the countryside” when graveyard beat reporters from Boston area newspapers landed Dad’s photo and his “words of wisdom” on several front pages. “The Boston Herald” headline read: “Flying Farmer Forced Down in Boston Harbor” and had a picture of Dad with his feet in the Coast Guard’s oven with his work boots drying on a radiator in the background.

All’s well that ends well, they say, and one might assume my father’s adventure was over. Not so. Dad intended to fly off Outer Brewster Island in the morning, refuel at the Beverly (Massachusetts) airport, and be home just in time to miss the morning milking.

While eating a “Coast-Guard-style” breakfast of bacon, eggs, homefries and coffee, Dad arranged to bum a ride out to the island with some of the “understrappers” who were delivering supplies to the lighthouse keeper on Little Brewster. They motored out with ten gallons of high-test gas when, halfway across Boston Harbor, a short-wave radio call came, ordering them to return to the station.

“Bring that man back here!” the message was. The boss Guardsman, “The Commander” nixed Dad’s easy exit. They brought him back.

“So The Commander, he tells me he’d been checking around and was talking with a ‘Mr. Hurst’ at Beverly Airport. And when I heard that name, I knew I was in trouble because he's one of the guys who came up to arrest me a couple months before. Not arrest me, just serve me with a summons to appear in court because I landed at their closed airport."

(... which is a whole separate story where Dad had delivered bags of cow manure to his brother Edward in Hyannis and stopped to pick up rake teeth that the tractor salesman left hanging on a fence at the airport. But the landing strip was closed for a stock car race. And this Mr. Hurst was the "man running around in short pants." He had emergency flashers on his truck and tried to block Dad's landing. But Dad "sideslipped her down there" onto grass beside the fence and took his teeth. They probably wouldn't have pursued Dad across the state line if he hadn't made fun of Mr. Hurst's pants.)

"So, The Commander told me he swore to protect and rescue people, not to help them commit the 'pure suicide' that Mr. Hurst claimed would happen if I tried to take off from where I had landed." Coast Guard staff had already begun considering ways to disassemble the Defender and bring it ashore to be transported by truck back to Epping.

No doubt, Dad, having been through bureaucratic wringers before, was feeling a sense of urgency, a need to get out of there before too many more experts were consulted. He told The Commander not to listen to "the man in short pants" because his opinion had been prejudiced by a previous "unfortunate incident." He asked to talk with the director of the Massachusetts State Aviation Board who had testified when Dad appeared before the judge in Beverly.

"We talked over the phone, the director, The Commander, and me. And the director said if I thought I could take off from there, then he knew I could, and it would be my own damn fault if I ended up in the drink. He told The Commander that he didn't have a problem with me trying."

So, The Commander wouldn't let any of his men assist, but looked the other way when Dad received offers for a ride to Outer Brewster. He went with a caretaker and other locals who wanted to see if he could do it. They helped Dad drag the Aeronca "down into a hole" on the farthest downwind corner of the bluff.

"Then I borrowed their shovel and sank chunks of wood to trig the front of each ski, just enough to hold me back while I got her wound up good." Dad was pretty sure he would make it into the air because he didn't have a full load — not much gas and it was just him. "But if she wasn't flying by the time the bluff ended, then we'd slip down about twenty feet onto the lower level and skidder along there. And, if that didn't work, then we'd have to water ski and hope for the best. If I sank in the cold water, I probably wouldn't last too long."

Fortunately, Dad's "bird" took flight right from the top without any skiddering, water skiing or sinking. Later, spectators said they were nearly blinded by propeller-wash sand.

Dad's homecoming was exciting. Word had spread enough to attract about 50 friends and neighbors to the crest of Red Oak Hill. He had set 10:30 as his ETA. So everyone gazed south across the gray wintery woods of downtown Epping toward the rolling hills of coastal Massachusetts. Each of us hoped to be the first to yell: "There he is!"

Eventually, someone spotted a little dot against the vast blue sky. Dad landed back at the farm, kissed Mom and told everyone: "There's no heroics here. Any fool can run out of gas."

Hero or not, he seemed to enjoy his adventure more than regret missing two milkings.

Uncle John Gilbert
Reverend & Barn
Builder

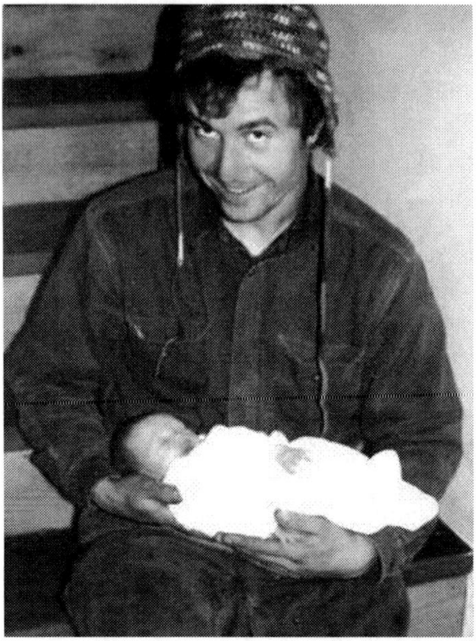

'Uncle Amos' (our younger son) with Rachel Elizabeth Clapp our first grandchild, born to Otis and Melanie October 10, 2009. They all live at 'Square One', our Epping woods.

Cindy and her horse 'Dan'. She is in costume for a 'Harvest Hunt' she organized to raise tax revenue for the farm.

PART THREE

Reduced to Rabbits

The way we are ...

23. RABBIT. RABBIT. WHO'S GOT THE RABBITS?

The Sisters' Refrain (please) ...
How can we be sure she'll stay gone
When we're not there all day long?
We'll yank out her tank.
She'll have us to thank.
In the end, we will see who was wrong.

* * *

Where there were cows, hay and oats
And wet manure floating in moats,
Now sit chairs of great size
And plumbing supplies,
And, zounds! We've even got boats.

There was the problem with the missing wrenches. Erica saw Phyllis and Beth drive away in a van and then the tools were GONE.

"They took all the box-ends and even the adjustable wrenches!" my youngest sister declared. What did I think we should do about THAT?

It was a darn good question!

What SHOULD we do? And how had the magnificent Applehurst Farm become a stage for squabbling over disappeared cats, runaway rabbits and misplaced wrenches? It felt like a divorce where attorneys are hired to fight for the toaster, file cabinet and cocker spaniel.

Does anyone really care about the tools and pets? Are the five younger Harvey Girls asking themselves: "Am I my sister's rabbit keeper?" and "May I please hold your box end?"

I doubt it.

They're just posturing to exercise power over each other, seeing who can be most indignant and stubborn, tending to lean on the fences that define each other's personal space and physical territory.

Part Three – Reduced To Rabbits

Indeed, a recent skirmish was over just that: an electric fence across the Night Pasture. Cindy and Erica had removed it. Phyllis put it back up. She said it was HER fence. They said it was in THEIR way.

Epping police were summoned.

Since when do we need police scratching their heads over our family affairs?!

While psychological taunting could be put down to bitchiness, recent blatant physical maneuvers over use of buildings and acres of land were unambiguous in their intent to get rid of the cows. There was nothing subtle about The Sisters' telling a plumber to deactivate the milking parlor the week after Phyllis was evicted.

My cow-loving sister went wild when she discovered what she considered to be HER space (the milk room, parlor and holding area) piled high with plumbing supplies. Warehouse crates containing ductwork, plastic pipe and insulation were stacked to the ceiling. Phyllis claims to have "gleefully" heaved the boxes out into the driveway, though I doubt that she was smiling at the time.

Next, The Sisters removed a several-thousand-dollar cooling tank from the milkroom – pulled it like a bad tooth, dragged it Up Back to the junkyard and draped a blue tarp over it.

When they boarded over the door, removed the water heater and trashed milk production records, the place was certified lactose free and Phyllis took the hint. She said she no longer felt welcome at the farm.

Initiative with consequences has always been 'The Harvey Way'. We were raised to "just go ahead and don't worry about what others think" because not everyone will ever approve and it is important to do what you think is right.

This worked okay while we were young and full of promise, loosely pulling together to 'farm'. There was a unity of purpose and, since we were all more-or-less volunteers, it seemed to me (at least) that we could confidently do our own 'things' to help out while being more-or-less appreciated for our efforts on the farm's behalf.

But now, as we have aged and become less tolerant, and have (basically) lost our senses of humor about farming, it's not so easy. Time to get serious.

Phyllis had reactivated the dairy operation in 1994 without consulting or gaining unanimous consent from the family. She had 'just gone ahead'. So, now –15 years later – a harsh financial spotlight has exposed the cows as losers, and Phyllis has been unceremoniously ousted. The Sisters have 'just gone ahead' and booted her out.

Playing by the same rule, Erica and Susie recently converted the farmhouse into three apartments. The Sisters 'just went ahead' and lay claim to the building and converted my bedroom into a bathroom.

Can't say I blame any of them for not clearing their intentions with the rest of us. We have grown up in each others' pockets with a critical dose of 'what's-mine-is-mine-and-what's-yours-could-be-mine-too-if-you-don't-shape-up' attitude.

It's a curse to crave approval from a family not inclined to grant it. We have yet to all agree on anything. Even when we make a point of coming together, some don't – like for family meetings and Christmas, or Erica's or my son's weddings, or birthdays, or putting up firewood. So there is little incentive to seek consensus – perhaps a consequence of our training to 'just go ahead' and don't worry what others think.

We were taught to 'Question Authority' long before it became a bumper sticker.

Besides, those who pay get to play.

Phyllis made deals and bought her own dairy equipment. Erica and husband Steve's pet cremation business funded most recent building projects – major investments like converting the farm house and building our parents a whole new more comfortable house and providing maintenance equipment and other supports.

You got the money, Honey, you GO!

Since removing Phyllis and the cows, Erica has begun orchestrating an overhaul of the vacated dairy facility. Last I heard, she bought a doughnut machine to attract more customers to the vegetable stand. It will take a while to recoup the investment, but it's none of my business. Mom seems well pleased and that's worth a lot.

Certainly, Erica's generosity is beyond the call of duty and out of my league. David and I don't earn enough money to support the farm and we don't expect it to support us.

Part Three – Reduced To Rabbits

Different from my stance, Phyllis thinks she owns a piece of the agriculture and would like to earn a living from it. Same as the rest of us, she is listed as an LLC shareholder, plus she holds a lease on the dairy facility and Night Pasture.

An objective reader might assume that Applehurst LLC was formed to encourage Phyllis's efforts. The farm's legal documents stress repeatedly the importance of preserving New Hampshire agriculture through growing crops and producing milk. The estate plan says farming 'shall' continue after our parents die.

But it doesn't seem to be headed that way, since the dairy aspect has been extinquished for economic reasons. Susie has already reined-in the checkbook, staunching the flow of cash that Dad tipped toward Phyllis's 'extra' farming expenses.

"It's not fair," Phyllis says.

"It's just business," says Susie.

Dad no longer mentions his "little rubber checkbook" and probably hopes these tensions will blow over into the 'forgotten' zone where other uneasy states of affair are trying to R.I.P – most notably my two brothers' relegation to a second line on the inheritance front. Neither Matthew nor Albert nor their four sons will carry on with the farm as Harvey men have done in the past. Indeed, all of my parents' grandchildren are intentionally left out of the Limited Liability Company. I paid my lawyer $400 to ask: 'Why?' and was told: "It is the way your parents want it."

Please understand that I've always been a women's rights advocate. I was quick to burn my bra in 1969, shunning eye shadow and high-heeled shoes and worked awhile for the Maine Women's Lobby, advocating women's liberty and equality. So it's not as if I'm hung up on preserving traditions of male dominance within society. Still, I lament that our bicentennial family farm has veered away from my brothers to favor the girls.

I suppose if my parents can have favorites, so can I. You will not be surprised to learn that Matthew has always been my hero. And it's not just because we saved each other's lives – though it helped.

I'll never forget how much he resembled John Wayne as he clambered, fell, and got up again, flailing in his rush across the rough field. He had jumped off the plow tractor to save me.

I like to think he had telepathically heard my cry "MAAATHEW!" as he disappeared like the setting sun over the hilltop horizon. I was on a separate (tricycle-style) John Deere, dragging a harrow to smooth overturned earth when the front tires dropped into a woodchuck hole at the same instant the large right wheel bounced onto a boulder and left me balancing on a half-rolled tractor.

"MAAATHEW!" I screamed as my eyes locked onto his shoulders and head fading from sight ... GONE! It felt like a miracle when he came running back. He pulled the tractor wheel down and declared: "There! Now we're even" – referring to the time I saved him from drowning in the manure pit.

The ties that bind!

I can't explain why my siblings have grown apart while remaining rooted in the same Epping soil – except to speculate that when the cows were terminated in 1987, there was no longer a reason to pull together as a family. When chores ended, so did our identity as the 'Harvey' team. We sacrificed our brotherly and sisterly love while turning our backs on farming.

As far as I was concerned, everything changed in 1987. It was time to seek mates and off-farm employment. My siblings walked different paths while circling the abandoned cow yard. I drove up Interstate 95 to raise our family in Maine for the next twelve years. My choice. No regrets. No apology. No farming aspirations, since I have never been willing to anchor myself to Applehurst Farm like my brothers and sisters.

Susie, Matthew, Phyllis, Albert, Cindy, Beth and Erica deserve immense credit. They each traveled some, but basically have never left our parents' side, remaining available to help on the farm while I have continued to spend year after year away.

In 2003, when David and I began wintering in New Zealand, we sold our (very nice) house in Maine and returned to Epping. The four of us came back to 'Camp David', our 30-acre woodland-swamp, and renamed it: 'Square One'.

My brothers and sisters have separate homes. Some of us have spares. Seven out of eight have built at least one house ourselves. I, for example, am on the 'Five Shack System', with our best dwelling in Riverton, New Zealand. Susie has at least four houses; Matthew two; Erica three; and Beth owns several mobile homes. Poor Phyllis, Albert and Cindy have only one house apiece – as far as I know.

We are a fortunate lot, neither to be envied nor pitied. And, considering the dramas of other families, we probably get along better than most -- rabbits, wrenches and relocatable fences notwithstanding.

Can't help coming back to George Orwell's description in "Animal Farm" where Chapter X provides insight:

"... And yet the animals never gave up hope. More, they never lost, even for an instant, their sense of honour and privilege in being members of Animal Farm. They were still the only farm in the whole county – in all England! – owned and operated by animals. Not one of them, not even the youngest, not even the newcomers who had been brought from farms ten or twenty miles away, ever ceased to marvel at that. And when they heard the gun booming and saw the green flag fluttering at the masthead, their hearts swelled with imperishable pride, and the talk turned always towards the old heroic days, the expulsion of Jones, the writing of the Seven Commandments, the great battles in which the human invaders had been defeated. None of the old dreams had been abandoned."

Later ...

"Benjamin felt a nose nuzzling at his shoulder. He looked round. It was Clover. Her old eyes looked dimmer than ever. Without saying anything, she tugged gently at his mane and led him round to the end of the big barn, where the Seven Commandments were written. For a minute or two they stood gazing at the tatted wall with its white lettering.

"My sight is failing," she said finally. "Even when I was young I could not have read what was written there. But it appears to me that that wall looks different. Are the Seven Commandments the same as they used to be, Benjamin?"

"For once Benjamin consented to break his rule, and he read out to her what was written on the wall. There was nothing there now except a single Commandment. It ran:

'ALL ANIMALS ARE EQUAL

BUT SOME ANIMALS ARE MORE EQUAL THAN OTHERS'"

Tow it long enough and it WILL start! Matthew pouring gas into carburetor of 'Harvey Farm Haying Crew' '57 International.

One of our five shacks: 'Beaver' cottage at Camp Menatoma, Readfield, Maine.

Our 'Cosy Dell' in Riverton, South Island, New Zealand.

Back to 'Square One' for summer in Epping.

The way we are ...

24. I DON'T KNOW JACK

The Sisters' Refrain (please) ...

We hate to seem viscous or mean
But if we were to come clean
It's failure to pay
Their taxes today
That makes us turn livid and scream.

Erica sputtered: "I told Albert about the wrenches. He said he wasn't surprised and didn't expect anything to come of it." The shop Up Back had been cleaned out. My youngest sister felt obliged to notify us all and kept asking: "What SHOULD we do?"

It took me nearly a week to respond and, in the end, I e-mailed her exactly what I had drafted as my initial reaction:

Hi Erica! Sorry about delay in responding to news of the shop being cleaned out.

My immediate reaction was that I agree with Albert. I'm not surprised and I don't think anything SHOULD come of it.

Didn't think you'd like my opinion. So let it sit a week. Now it still seems best I can do.

No. I don't know what to make of tool removal. Not sure I even care, though, since Phyllis and Beth have been proprietors of the shop for the past 12 years and probably feel they own the tools.

If I worked in a place that long without sharing, I'd feel territorial, too.

Wouldn't you?

Now, since they have both been kicked off the island, it would be reasonable to assume they are setting up a mechanic shop elsewhere. Probably at Phyllis's. So why wouldn't they feel entitled to take the tools?

In the scope of things, I can't get too worked up over the box-end or even the adjustable wrenches.

Love, carol

And before that, there was the problem with the rabbits. I got wind of that because I was beside Mom's tomato tunnel house weeding onions when Cindy and Erica roared into the upper driveway in their hefty pickup trucks.

"Those idiots!" exclaimed Cindy, who is 10 years younger than I am. She is the middle sister among the three who are trying to run the farm and save Phyllis and Beth from themselves. The (3) Sisters share a family counselor named "Jack." He helps them deal with stress and declared: "It's time for the circus to leave town!" (meaning, I guess, to get rid of the cows).

I don't know Jack. Just Susie, Erica and Cindy know Jack.

"Don't bother looking!" Erica called out. "The rabbits are GONE." That's when Cindy commented on idiocy and I walked toward them from the garden.

They said Beth had left town and her rabbits weren't being cared for, so they took them (Beth's rabbits) and were "cleaning up the mess," separating the bunnies so there wouldn't be incest, "throwing out the dead ones," hosing the cages and such. They had even arranged to give the rabbits back to the person Beth bought them from six months ago for $300. Eight rabbits, maybe. I don't really know.

But before the giving-them-back part, The Sisters had hauled the hutches from Beth's house, two miles away, and put them in a shed Up Back.

Two weeks prior to this unauthorized relocation, Beth left town after Susie told her to "just leave." Not much was said about Beth's departure, but Matthew drove to Amesbury, Massachusetts and found her living on a farm with her boyfriend Dave, who nobody but Beth seems to like.

It seems that it was Phyllis who moved the rabbits, again – from Up Back to her place – which would've been easy enough to do without anyone noticing, since her 'Edgerly' farm abuts the pasture and is separated by a gate near the cemetery.

"So. All right! So what?!! I took the rabbits, OKAY?! Guilty as charged!" Phyllis laughed later at the crime scene. "Big furry deal!"

She said Beth had phoned (from Amesbury) and asked her to reclaim them after being tipped off that the rabbits were headed back to their original owner.

"Those Girls had no right to go poking around Beth's property like that!" Phyllis declared. "And it certainly wasn't their place to be giving Beth's animals away. If they're so territorial, they should at least keep out of OUR territory!"

Phyllis went on to tell me about Beth's cats, also left behind, a litter of good-sized kittens, five of them, that she (Phyllis) had been asked to give away.

"I hired Olivia (our Australian niece) to catch them and she got two right away, then another," Phyllis said. Arrangements had been made with new owners coming to pick them up from a cage at the vegetable stand.

"But, overnight, the kits disappeared," Phyllis sighed, raising her arms. "Mommy said Erica had stopped by to 'take care of them'" ... which is code for having them euthanized by a veterinarian, since Erica's business is pet cremation and she has agreements with a dozen vets throughout southeast New Hampshire.

"I expressed my concern that the kittens were already spoken for," Phyllis said. "But it was as if Mommy's hearing aid suddenly failed and she changed the subject. I was left feeling sorry for the kittens, wondering HOW Erica got into MY cat project ... without permission. And I'm stuck in the middle to explain to the new owners WHERE are the kittens? This meddling has to stop! Whether it's cats or friends, the problem is: 'lack of respect for one's property and privacy!'"

Back at the garden with the pickups, Cindy moaned: "NOW what are we gonna do? I've spent a week arranging to give them back." (That would be the rabbits.) "NOW what will Mary think? She said they are valuable and she gave Beth a good deal on them. She expected they'd be taken care of, but now she wants them back. Mary will be mad."

During summers while David and I are in Epping, we join our son Amos's 'Tuesday Supper at Grammy's' club, where our kids and usually Christa and Abbie (Beth and Cindy's daughters) eat dinner at my parents' house, 'Grammy and Grampy's'.

Tuesday was the next night after my crash course on rabbit and cat troubles. When I arrived early at the parents', my mother started in about how disappointed she is, how "disappointed we ALL are" with Phyllis and Beth and the rabbits. "Mary is worried about the rabbits."

I thought: "Yeah! And what about the kittens that had been 'taken care of'?!" But remained silent, pretending to know nothing about the situation because nobody expects me to know anything anyway.

"That's too bad Mary is worried about the rabbits," I affirmed without asking: "WHAT rabbits? or WHO's Mary?" ...

Continuing my silent dialog: "This is TOO weird! Let's just forget it even happened."

While home for summer, we eat supper with my parents on Tuesdays – sometimes at our 'Square One' in the woods. Picture of our 16-20' brick house before recent addition for arrival of Granddaughter Rachel Elizabeth, born 10/10/09 to Otis and Melanie. Our sons were born in the brick house: Otis 5/25/81; Amos 4/28/83. We built it in 1979 after a loose-itinerary 15-month-long trip that included Alaska, Hawaii and New Zealand.

Cousin Linda Peterson at hilltop horse event run by Cindy.

Cynthia Ellen Harvey, Epping mail carrier.

David and I have grown up to be migrant farm workers:

Apple picking in Central Otago, New Zealand: Carey, Wilhelm, me, David, Benny and Joseph. Below: David, Amos & Otis – masonry.

Part Three – Reduced To Rabbits

The way we are ...

25. FORGOTTEN

The Sisters' Refrain (please) ...

> ***Of all the things, and they're a lot***
> ***That tie your heart in a knot***
> ***Is the feeling that we're***
> ***Held hostage by fear,***
> ***And, forgetting all that we've got.***

Isn't that what they say? "Some things are best forgotten."

It certainly works for us Harveys! We're very good at forgetting. We do it all the time. Collectively. All ten of us, our parents and eight grown kids. It's as if a lot of things never happened:

Good things:

- Like Grammy Peterson driving with me cross-country in 'Charlie', my creatively-painted '66 VW bus, and

- Dad being featured in 'Portraits of America' on Ted Turner's cable network as an appropriately outstanding New Hampshire Yankee, and

- Mom starting Epping's school lunch program in the Grange Hall and now, 50 years later, she's leading a campaign to restore the building as an historic site, and

- Susie and me traveling to Seattle for summer adventures, so clueless that we were surprised to see Mount Rainier as our plane circled for landing; I worked for a poverty program while she nannied for the oil-company Bardahl Family, and

- Matthew inventing 'The Harvey Drawback' for hauling silage trucks and continuing to help us all through personal hardship and building projects, and

- Phyllis becoming a private pilot, speaking French, spending a year in France, and being attentive to her many friends, and

- Albert being an accomplished SCUBA diver, running his own garage and inventing a rotary engine while raising three professional sons even though he divorced when the youngest was an infant, and

- Cindy playing basketball in Las Vegas and becoming Epping's best rural mail carrier ever, and

- Beth and Erica traveling through Europe and being in London for Charles and Diana's wedding, and both being outstanding high school athletes.

They used to be best friends, Beth and Erica. Susie and me, too. We've traveled together, just Susie and me, and still occasionally help each other out.

And we forget bad things feeding our family's undercurrent of low-grade resentment:

- Like Dad borrowing too much money and driving lawyers nuts with politics, and

- Mom always worrying about us 'kids' not getting along, and

- Us 'kids' not getting along for reasons we can't remember – except for me who remembers burning the brownies one too many times so Mom put Susie in charge.

We try to forget, which is okay to a point.

All memories mellow with age. Good memories fade pleasantly with no ill effect and sometimes sweeten. But I'm not so sure about dismissing the bad stuff, too. Is it healthy to repress resentment over things like our parents dividing 250 acres in Nottingham between we three oldest offspring and not the other five, and … ?

Well! How can there be no hard feelings? It's nobody's fault. Sometimes it is difficult to anticipate the future. Not even our parents are perfect. Still, it would be hard to deny that inevitable consequences have resulted from certain initiatives.

I'm sorry that Albert has received just two acres to split between his three sons – three of my parents' grandsons (who have now produced three great-grandchildren). In contrast, I was given 80 of the Nottingham acres, paid taxes on it for ten years, sold it to fund our move to Maine. That's it! Cashed in my chips and left the party. But at least I HAD chips and a seat at the table with the option of walking away, for which I am eternally grateful.

Is it good to pretend that Phyllis wasn't evicted? Or that Beth isn't angry because nobody likes her boyfriend? Or that the farm's attorney has never told us how our parents' estate plan is supposed to work?

These unmentioned conditions trouble me.

Part Three – Reduced To Rabbits

We Harveys celebrate and cherish memories of picnics at Adams Point and the annual Christmas gift swap and dinner. But we also suffer veiled distress over incidents that nobody acknowledges or articulates, like arguments over tractors, rakes, friends and relocated fences, wrenches, cats and rabbits. Plus there are larger, perhaps valid, concerns over welfare and safety, struggles over territory and control, elbowing for a better spot.

All to be expected, considering how, when and where were brought up. We are neither better nor worse than other families.

We have to remember and appreciate that Dad's relationship with real estate is different from the way most people think about owning land. He is not a speculator or a person who just needs a place to live. For my father, land around Red Oak Hill is his heritage and serves as a tribute to his life – evidence that he stood steward at the door and retained ownership of the Harvey home place.

If land was your life, you might not like to be told what to do with it, either.

I figure our parents didn't have as much difficulty as we do getting along with siblings. Their generation followed more traditional rules. Dad inherited the farm. Without fanfare, his sisters married and went away. Mom and her sisters, Ann and Alberta, also married and left Brookvale Farm. Apparently, it was no big deal. Their brother (Uncle Philip) got the place.

But for us Harveys, no such movement. None of us eight offspring have graciously bowed out to forfeit interest in the farm. Our parents provide no role model for this stage of our lives. They stopped living with their siblings while still in their 20s. We of the next generation are in our 40s to 60s. We make up our own rules, not always sanctioned by the Golden One.

Critical opinions from imperfect encounters ebb and flow throughout the family. Like the tide, it's hard to escape the waves of angst that, without warning, swell to buffet our charming rural facade.

This is not to deny our privileged childhoods or throw cold water on our integrity within the community. Our character as Harveys is generally impeccable. Or at least above average. While some of our neighbors have swapped spouses, sold drugs, embezzled money, stabbed, shot and bitten off each others noses and set fire to corpses on the front lawn, we have resisted such behavior thus far.

Still, we can't escape ourselves. ALL Epping Harveys have settled right there within walking distance of home – my parents, we eight 'kids', plus spouses and grandchildren with a slew of aunts, uncles and first- and second-cousins not far away – probably 60 direct-line relatives in Epping or just over the line in Nottingham, Fremont, Exeter or Durham.

I am grateful to the older generations, but also believe in the youth. The kids are ready and able to take over. I am eager to give Otis and Amos everything we've got so the goods can be shunted ahead to the grandchildren.

We Harvey-Petersons have breathed the same air, waved at each other daily on the same roads, and shared the same life. Our stories are similar, but our perspectives are not. This becomes apparent during each healing memorial service for a beloved family member. We all listen and love each other again, reconnecting with what we've forgotten – that bond special only to family.

Let bygones be bygones!

Oh, we forgot that part. The part about forgiving.

Some families are the other way around. They forgive, but never forget. Harveys are better at forgetting, so feel no need to forgive.

Dad and his "Little Sisters" Agnes Harvey Wilkos of East Glastonbury, Connecticut, born October 14, 1923, and Edna Frances Harvey Woodward of Durham, New Hampshire, born March 18, 1922. I don't have pictures of my Harvey grandparents (archived at Epping Historical Society). Matthew Harvey was born in 1870 and Alice Webster Harvey on June 27, 1882.

Part Three – Reduced To Rabbits

Historically speaking ...

26. N.H. DAIRY INDUSTRY - BLOOD ON OUR HANDS

The Sisters' Refrain (please) ...

> ***I like it. It's good to be King.***
> ***The sound of it has a nice ring.***
> ***Milk makes no cash.***
> ***I'll settle their hash.***
> ***So sell all the cows: Ching! Ching!***

When the Harveys went out of the dairy business in 1987, government experts advised that we could bid "zero" in the 'Buy-out' and do better than continuing with 100 milkers. That's what Cooperative Extension Service consultants said then. And now, in 2008, the same federal agency reports that: "During the past decade, New Hampshire dairy farmers have been selling milk at less than their production costs," and asks: "How can they continue to do so?"

We didn't. And now Phyllis has been stopped because our sisters declared her to be a waste of money and said the farm can no longer afford to sponsor a dairy herd.

But, you must understand. It's not just us! It wasn't our fault in 1987. And it's not Phyllis's fault now that her Holsteins can't produce enough milk to pay for equipment, fuel, maintenance and taxes on the barns and hay fields. Indeed, it's not even our parents' and The Sisters' fault for evicting the cows for economic reasons. They feel they have made a sensible business decision.

Four New Hampshire farms went out of dairying just last month (131, down to 127) with plans to cooperatively terminate two more soon. They say each dairy cow loses $3 a day.

So, Phyllis was shoved out the door just in time. She was just a peep-squeak operation, anyway!

It's just a farm. Not a matter of life or death.

Or is it?

Some people study medicine. Others become doctors or join gangs or the police department to learn about death. I suppose it's what religion is all about: living, then dying while hoping to be resurrected or cured. People learn about death in various ways.

Farm kids learn about it in their front yards when a cattle truck backs in to haul livestock away for slaughter. Or when the family dog is run over by a tractor.

Or, as with the demise of one of my 'Softpaw' cats, a cow accidentally steps on its head and starts it spasmodically flopping in the sawdust.

"Leave it alone!" Leslie, the hired man, yelled.

But there was no stopping my lunge to grab the pet that had been so sweetly walking toward me. She dug scratches into my arm, neck and stomach during her death throes. The bathtub water turned blood-red as I sobbed and pledged to pick the scabs to make scars so I'd never forget my black-and-white cat's passing.

That pledge and others regarding empty stalls, hen yards and pig pens were forgotten as soon as another 'Softpaw' was born and the abandoned living spaces were refilled with replacement livestock. It was the beginning of my understanding that women and children didn't have a lot of say when it came to life and death on the farm. I mean, remember the hired man drowning cats in the milk tank? Dad said he didn't expect us to get "all upset" over it. Mom had been mad and wanted him to fire Russell because he lacked common sense.

But Dad said he couldn't fire a neighbor, and it didn't matter what we kids thought. We just had to grow up.

It isn't that Dad was (or is) a heartless cat killer or rushes to ship cows for slaughter. He isn't and he doesn't. He was raised to be kind to children and animals. He often stopped to scratch a calf's topknot or rub a cow behind her ears. Our dog, Prince, liked Dad best. And when we got a horse, it was Dad who carried carrot treats in his pocket for it. He often referred to Grampy Peterson's "way with animals" and seemed to want to emulate his father-in-law's ability to work with livestock.

Indeed, other Epping farmers joke about how soft Dad is. They recall the year they decided to team up like the agricultural "coops" they'd heard about. We all got along well enough, "cooperating" with the McPhees and Allens, sharing our talent, muscle and an occasional side-delivery rake in a united effort to succeed with dairy farming. But when it came to culling cows, each farm had been on its own when milk prices dropped. Farmers killed cows to save feed, each farm negotiating its best deal with several livestock dealers who circled each week like vultures.

The year of cooperation was especially bad because, not only was the price of milk low, but import of Argentine beef made domestic meat nearly worthless as well. New England farmers were more stuck than usual. So, we Harveys, led by Matthew, joined with McPhees and Allens to set up a mobile slaughter house and hired our neighbor 'Little Roley' Letourneau to cut up the sacrificial cows.

Peer pressure forced Dad to cooperate. Left to his own devices, he would have called a cattle dealer and accepted a low price or he might have given the doomed milkers to a needy family. BUT ... if he was the farmer he claimed to be ("the biggest farmer in Epping"), then he had to go along with the younger guys.

"Bleed 'em and hang 'em high!" crowed Paul McPhee, the rowdiest of the bunch.

"Let the slaughter begin!" added Bruce Allen, who was equally up to the task.

The plan was to walk the cows, one at a time, into Bruce's ton-and-a-half stock truck. Little Roley would shoot the beast in the head and cut her throat. "Before she knew what hit her," the farmers would chain its rear legs to a splay board and hoist the carcass to dangle upside-down from a raised tractor bucket.

It worked slick. But Dad had deliberately missed practice sessions. He had let Matthew go help at McPhees' and Allens'.

So Dad didn't know what to expect when the truck backed up to our own barn.

"You should've seen your father crawl over that cow as she fell!" taunted Paul McPhee. "He never expected Roley to shoot her so fast. I thought he'd fall out of the truck on his face!"

Harr! Harr! Harr!! Big joke on Dad.

They didn't 'cooperate' on much after that. And, as far as I know, my father still has never been able to kill anything himself – not with his own hands – evidence of a chink in the armor required to manage a farm where milk production is the top priority. Animals detracting from that goal are killed. Though Mom still doesn't seem to mind whacking off chickens' heads, it was very difficult for our family to send livestock for slaughter.

The bigger farming gets, the more ruthless it becomes. Some say the U.S. Department of Agriculture hasn't helped family farms at all by constantly pushing for economies of scale. Indeed, many of the still-standing dairies are much larger than our Applehurst Farm ever was (with 100 milkers). Some of these surviving producers are too deep in debt to quit, while others enjoy comfortable financial cushions, often underwritten by public or private endowment. Some farms intentionally operate at a loss for corporate tax write offs.

Without profit, income tax has never been one of our worries.

In 1986-87, the Federal Whole Herd Termination Program shut down 14,000 dairy farms nationwide. We slaughtered our herd along with 58 other Granite State families. The government spent $2 billion to eliminate 10 percent of the nation's milk supply to reduce subsidy for surplus dairy products. Half a million cows, heifers and calves were "terminated," slaughtered or shipped out of the country.

But it didn't work.

In retrospect, the Department of Agriculture acknowledged that the Buy-out did little to raise the milk price paid to farmers and failed to significantly reduce surpluses. At first, milk production appeared to be reduced by 8.7 percent. But, upon closer examination, it actually increased 1.2 percent after 1987.

Why?

Because, while 14,000 farms were shut down for five years, the remaining 125,000 dairy operators took advantage of the situation. They, being good capitalists, saw what was going on and doubled their herd replacement capacity from 25 heifers per 100 milking cows to 50 heifers to bump up production per 100 milkers. Not much cooperation for the overall benefit of the industry, just a grab for informed short-term profit.

Part Three – Reduced To Rabbits

The Buy-out made little difference to the nation, yet our Harvey Family's way of life would never be the same without the cows.

New Hampshire now has fewer than 10,000 milkers on 127 farms, down from over 115,000 head on 22,500 dairy farms in 1900. Since World War II, 30 million American dairymen have chosen not to go into debt to continue an agricultural life style. They quit farming.

And what does it matter, really, if kids don't understand that milk comes from cows? Let them think it's mixed up in the back room of a supermarket or bottled by Coca Cola. At present, in 2009, Washington lobbyists are pushing for another 'buy-out'. Milk consumption is down while production surges and the price paid to farmers sucks canal water. The dairy industry expects to be 'bought out' (is that the same as 'bribed'?) to kill 250,000 to 300,000 perfectly good cows.

Too much milk, again.

Too little love, as usual.

"Up Back" and crest of Red Oak Hill where Dad bulldozed and we picked rocks.

Today's scenery: Fernald machinery working the crest of hill above our nieces' little house beside farm pond installed through government grant. Photo was taken from Red Oak Hill road beside 'Ice Cream' sign pointed toward 'The Milk Room'.

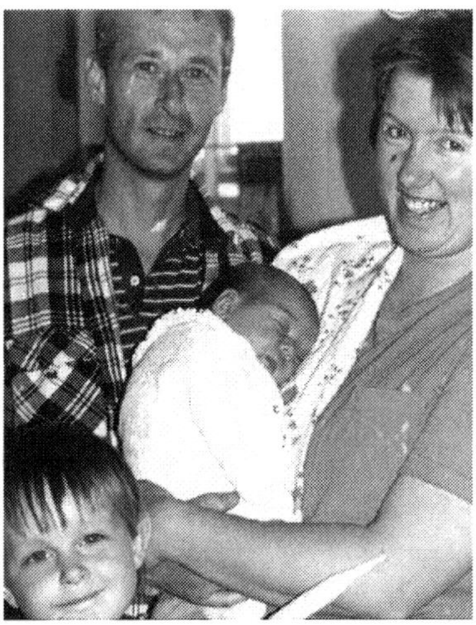

The Weaver Family 18 years ago: Sam, Beth, Johnathan and Baby Christa.

'Tuesday Night at Grammy & Grampy's': Mom, Me, David, Dad, Rachel (6 mos) and Melanie Fike Clapp.

Generations of kids have played 'Up Back' in the junkyard. One year, Matthew and Albert's kids published a 'Red Oak Hill Report' that included editorials on how they were paid in 'PTs' – Play Time hours where they were allowed to sit on immobile tractors while making 'motor sounds'.

Mulching in rainy York, Maine as 'Full Moon Gardening':Jennifer Marion and me with our boss Doreen Casey.

The way we were …

27. RUSTLERS

Susie, Matthew and I complained about finishing our barn chores before the mid-October sun rose.

"Still dark!" Susie said as I slid the horse barn door open so we could step out into the day. Beginning at six, we had fed and 'cleaned out' 24 calves and two horses, slopped two pigs and collected eggs from 50 chickens. The three of us often disagreed about the division of labor in our shared workload, but after an hour-and-a-half of exertion we usually emerged through the door in relaxed good spirits.

"Could be worse," I said on that Saturday morning. "We could've broken the light bulb. We could be Sulloses!" It was a joke.

Matthew had recently gone with Dad to the Sullos farm in West Epping. He returned with eyewitness confirmation of stories we had heard about Epping's fourth dairy farmer, the herdsman who seldom cooperated with the rest of us, the shell-shocked war veteran who never bought a new tractor and who refused to compare milk production figures.

Jim Sullos rarely left his farm and when he did, everyone commented that he smelled of cow manure. Refusing to modernize, he continued to milk by hand. It was said he wouldn't spend a nickel to see an earthquake and certainly wouldn't hire help. Dad asserted that Jim was clever, though, because he saved his money instead of spending it on "frivolous" improvements, like we did.

One story was that Mr. Sullos was so thrifty that he had only one light bulb in the barn.

"It's true!" Matthew said.

Dad had taken our eight-year-old brother "to see how the other half lives" because word was out that Jim Sullos had "walking pneumonia." Paul McPhee and Bruce Allen had asked Dad to go talk to the "old coot" and arrange a schedule for them to go help with milking and chores. The young men said Dad was the only person Jim ever listened to (probably because Dad was the only person patient enough to listen to Jim's war stories).

"We didn't think anyone was around," Matthew said. "The barn was dark and there was no compressor or anything running," observed my brother. Still, he and Dad knew the smelly bent-over farmer must be there somewhere because they had checked at the house first, where Mrs. Sullos said "James" was milking.

Besides, Dad knew what to expect. He just wanted Matthew to appreciate the 'things-could-be-worse' aspect of dairy farming.

"But we went in there anyway, where it was REALLY dark, and Dad tripped over a sill in the milk room. Then we ran into the cows tied up in a row coughing and rattling their chains. All we could see was their rear ends where the light came through the windows. Dad hollered: 'Jimmy! You in here?' and Mr. Sullos he hollered back: 'Who is it?' and around the corner there was a light."

Matthew was impressed that Mr. Sullos carried a 75-watt light bulb with him to use wherever he needed to see. He screwed it into a socket, then unscrewed it to screw in again where he went next.

We always had the Sullos model for reference in considering tradeoffs for what it would take to save, rather than spend, money to be 'progressive'. It's safe to assume that we Harveys wouldn't (and didn't) choose to turn a profit if it meant coughing and wheezing alone in the barn, light bulb in hand, faced with the rear ends of 20 boney Holsteins. We would rather borrow money from the Production Credit Service.

Back on Red Oak Hill, we enjoyed the weekend. No school. Susie, Matthew, Phyllis and I had spent two hours watching television while eating bacon, scrambled eggs and toast for breakfast. Leslie, our burly hired man who could lift a 70-pound jug of milk on a rope clenched in his teeth, hollered for us through the door. We three oldest climbed into the back of a pickup and headed for Nottingham.

The plan was to help drive 20 heifers home for the winter. After capturing them, Dad would meet us in the small corner field. He would then lead the rambunctious mob two miles down the dirt road to our farm. Dad was good at calling the cows and shaking a bucket of grain just enough to keep them interested and headed in the right direction, but not so interested that they'd steal the bucket from him.

"Something's not right," Leslie said when half the yearlings were missing from the pasture where he had dumped eight bales of hay for them on Wednesday. He directed Matthew, Susie and me to scoot through the woods to a nearby hayfield while he drove the ten heifers back to the corner field where we would rendezvous with Dad.

"Looks like they cleaned up the hay and went looking for more," Leslie said. "You kids, if you find them, chase them toward the road and we'll bring 'em back down to the corner."

The ground crunched beneath earthtones of fallen foliage. Only oak and beech trees still fluttered their brown and golden-bronze leaves as we walked among pine and hardwoods. Crows squawked and scolded as we swung our sticks single-file along the cows' woodland path. All of our pastures are irregular in shape, separated by patches of forest regrowth that continually spreads inward along the edges. The cows are never far from tree cover. We stopped to stomp smoke-bomb mushrooms and whacked spent pods off a milkweed patch as we entered the field.

The black-and-white heifers were there all right, grazing in mellow sunlight at the lower end of the field. We three cowpokes loped our imaginary horses down the slope and made quick work of delivering the wanderers to Leslie and Dad who waited at the roadside gate.

But something still wasn't right.

"There's one missing," the men agreed, though they weren't sure which one. They decided we'd better all walk the fences and look for tracks.

Giving Susie and me a nod, Dad said: "OK. You two start here, one each direction until you come 'round and meet over there somewhere. Yell if you find where they've been gettin' out. Matthew, you come with us. We'll check the pasture."

I strolled along the uphill edge of the field scanning two strands of barbed-wire stapled to crooked fence posts inside the stone wall. We Harveys weren't the best fencers. The cows and neighbors knew all about it. Our livestock were forever getting loose, not all at once, but a cow or two here or there. One summer, we paid an insurance settlement to Fran Allen, our life-long neighbor (Bruce's mother) for damage to her flower garden.

Part Three – Reduced To Rabbits

Dad was not above shipping a beast for slaughter if it refused to respect our best efforts to contain it. Harsh as that may seem, it was in line with what Paul McPhee did one year. He declared hunting season on a Charolais heifer that had turned wild.

"Whoever shoots her can have her!" Paul announced when the white cow refused to come in for the winter. The beef animal took to the woods and ran with a herd of white-tailed deer. Led by McPhee, the men stalked that cow for nearly a week. I remember Matthew's lack of enthusiasm for the hunt. He carried his rifle in its bag like an attaché case. Said he didn't want it to rust. The cow ended up in the local butcher's home freezer, shot in an unsportsman manner by his son Roley INSIDE our barn!

Susie and I circumnavigated the field. No sign of a break in the fence. No broken posts or fur on the barbed wire. I turned to see if Susie was still walking along her section of fence. She was, so I kicked the corner post and headed downhill following the wire as it turned from the stone wall.

Just over a knoll, barely out of sight from the road, I nearly spewed breakfast.

Intestines had spilled from what was left of 'LuLu', the missing heifer. I recognized her immediately because she was all there except for her hind quarters, which, apparently, had been hacked from her body with a knife and what? An ax? to smash through her hip bones.

I took a deep breath and yelled "H E L P!!" as loud as I could while scrambling back up the hill away from the gore among the trees. The faster I ran, the faster I felt like running toward Susie who had emerged from shadows in the lower field. I was horrified and just ran, not stopping to explain. My sister walked, rather than ran, forward. She said later that she thought it was a joke or that I was being chased by bees. When she saw my tears and the way I couldn't talk straight she knew it was no joke. I sat down abruptly and told Susie that LuLu had been murdered.

"Go get Dad!" I ordered. "Yell!"

My blonde sister was off like a shot, screaming: "D A D D Y! COME QUICK!!"

"The BAAHSTIDS!" Dad growled several times as we approached LuLu's remains, all that was left by an obviously inept butcher, a poacher who probably wasn't alone. Two full bottles of Budweiser had been left behind in a cardboard six-pack carrier beside a clump of ferns. Three empties had been tossed.

"Must've been in a hurry. They forgot their beer," Leslie said.

"Stay here and don't touch anything," Dad said to Susie, Matthew and me. We were happy to keep our distance. He told Leslie to walk back and get the truck. Then Dad did the weirdest thing. He sat down right there in the field and took off his work boots. The ground was COLD after several weeks of overnight frosts and Dad was rarely seen in his stocking feet. Certainly not outdoors. He explained that he would look for tracks and didn't want his to be confused with the poachers'.

The three of us stood in the field, close to each other and close enough to the kill site to ask Dad: "What happened?" He walked right up to the carcass and ran his hand along LuLu's neck and shoulder.

"She's been dead a while," Dad said. "A full day, I'd say. Pretty stiff, all right." What bothered our father most, and us, too, once we knew, was the waste. "Bad enough they killed her," he said. "But there's no excuse for such a hack job! If they didn't have time to skin her, they could've at least taken the whole thing."

Dad found plenty of tracks and returned to our sunny spot to remove his wet socks and put his boots back on. He guessed that three men had committed the crime: two in work boots and one in motorcycle boots with a distinctive heel. A trail of blood indicated that LuLu's hairy hind end had been carried by two men a short distance through the woods to the entrance of an abandoned logging road, not far from Thompsons' house.

We kids banged around in the back of the pick up as Leslie roared over ledges of the poorly-maintained Ledge Farm Road to 'Liar's Paradise', the only store in Nottingham, north of Epping, the town where our heifer had been slain.

Long story abridged, a sheriff's deputy was summoned and met us an hour later back at the guts. He studied the scene and agreed with Dad's assessment before telling us to go home because he could handle it from there. But Dad and I went with the deputy to the neighbors' – Thompsons, sending Susie and Matthew home with Leslie.

Part Three – Reduced To Rabbits

The Thompsons said, yes, Mary had noticed a blue car with Massachusetts plates and "three big guys" coming out the logging road and past their house on Thursday afternoon. But she didn't think anything of it because hunters were in and out of there all the time – not that deer season was open yet.

"But, ya know, hunters are always checkin' 'round ahead of time, gettin' ready, ya know, for the hunt," Mary told Dad and the deputy. Anyway, what was she supposed to do? "Report it to the police? What good would THAT do?!"

No good at all, we decided after burying LuLu – head, heart, hide, furry little pubescent udder, tiny teats, severed tail, and almost all. We never heard another word about it.

Around that time, two barns burned: one of ours in Nottingham and one of Uncle Ed and Aunt Ann's across the road from their house. Both piles of charred timber were investigated and found to have been ignited from multiple "hot" spots. Subsequent thefts and vandalized heavy equipment have similarly reeked of criminal intent. But, not once, have the law enforcers come through for us.

Local police appear to be recruited from a nonagricultural gene pool and don't understand that farmers suffer as much as people in subdivisions. The last responsive cop was Eddie Lavoie, back when he was our one-man police force. As soon as uniforms and cruisers came to town, it was out the window with follow-through justice.

As recently as last summer, I looked out one morning to see Epping police sitting in our driveway talking on their radio. They said they were looking for catalytic converters that had been stolen from a Stratham car dealership.

These cops had our son's license plate number. Oh yeah?! Just happened I knew all about it!

Amos had mentioned that he and his girlfriend stopped to look at a car on the front line at 'Stratham Porsche-Audi' and had been pulled over by a cruiser because they didn't look like they could've afforded to buy the car.

Amos said he had been "economically profiled" and wasn't happy about it. So here were our own Epping police pursuing non-criminals on behalf of an out-of-town business. They drove half a mile into the woods for no good reason, apparently unaware of the fact that my family would never steal catalytic converters or anything else.

I let them go.

But Matthew intercepted the officers as they were leaving. He had been lying under his chipper, greasing it, as the unmarked car passed through the stump dump.

"So! You guys look like detectives," my brother began, leaning into their black car. "Are you lost?" The police explained about the catalytic converters and wondered if he had seen any.

"No. Can't say I have," Matthew said. "I thought maybe you were here to say how my case is coming along." His equipment had been vandalized and reported to police a month earlier. Two-thousand-dollars-worth of damage to a truck transmission and glass smashed on a loader.

"What case is that?" they asked.

Without comment, Matthew turned and walked away with his grease gun.

Albert Adams Harvey

Otis and Amos in alfalfa. They called Grampy's pushed-aside boulders 'The Polar Caves' (at far end of field).

The way we were ...

28. SNAKE IN THE GRASS

Susie and I are walking out across the night pasture in our bare feet. We are eight and six years old. It's time to bring the cows in for the afternoon milking. We carry our favorite sticks to hit them. Another stinkin' hot day, but, hopefully, no thunderstorms. We need to find the herd in the lower Plumer field, beyond Nottingham Square Road. They will probably bc lying in shade.

A thrush scratches for bugs in a dry cow flap as we enter a pine grove, headed for the underpass. Startled sparrows jump into flight from blackberry bushes along the stone wall. A chunky brown-gray woodchuck butt flashes among the rocks.

"They've been having a zoo party without us," Susie says.

An indignant chipmunk scrambles to chide us from a stump. Chip!-Chip!-Chip! It chitters at us, alerting a Blue Jay who squawks from a limb.

"Ahhh. Shut up! It's just us," I say.

"Yeah. Shut up!" Susie mimics.

She is following me along a dusty path where the cows have trampled grass to death and scuffed aside pine needles and cones. The air smells like baked hay and dirt. Our feet are grimy gray up over our ankles. We don't bother squealing "Eeeooo!!" when we step in manure. Mom says it will make our feet grow bigger. A stone-lined tunnel under the road connects our Night Pasture with the Plumer fields. Underfoot, the underpass oozes cold black manure because the cows love to loiter under the road where it is cool. They crap into a stream that flows through into a pond.

Susie and I scurry through the dark weeping tunnel and step out into glaring sunshine. It reflects off the glassy surface of the nearby watering hole. We stand squinting among sedge grass tussocks. Every fall, Dad digs the pond deeper with his bucket tractor. He says not to worry about the cows slipping down the steep banks into water over their heads. He says they can swim. "If they slip, they'll just swim to the other side and walk out," Dad says.

I'd like to see that!

My little sister and I will use our sticks to make the herd move past this pond and enter the underpass. Our cows would prefer skipping milking altogether to loll in this low-lying oasis. Once we get the herd this far, Leslie, the hired man, will come over from the barn to help. He can't milk if he doesn't have cows.

Susie and I dash up a steep embankment through more pines to reach the closely-cropped late-summer pasture with its dark green patches where the cows have deposited nutrients to spawn fresh grass. The Plumer fields roll in a series of little hills, up and down, just enough to hide a cow from short people. We must conduct a thorough search. Susie strikes out for one distant corner of the field and I for another, along the road. Cow paths become less distinct in the grazing area. The ground is warm and short grass tickles the soles of my feet. If there's a mushroom puff ball, I stomp it. We call them "smoke bombs." Upright Scotch thistles are easy enough to avoid, but low flat prickers and sharp rocks are an ever-present danger as Susie and I stride confidently toward our corners, cow cudgels in hand.

Crickets chirp as robins tear into red sumac clusters growing along the road. The sky is clear pale blue. Good hay weather. No billowing clouds to rumble with thunder and dump rain onto Dad's new-mown crop. Bales are way too heavy when the hay gets wet. We have to store wet bales separate from the good stuff and spread it out, not pile the bales too tight, because Dad says they could "spontaneously combust" and burn our barn down. It happens, he says. Plum Ladd's sister's barn burned because of spontaneous combustion.

"Lionel" the mailman toots at me as he drives by. His nephew, Paul, is in my class. We have to go back soon. Yuuck! I hate school. I'm not finding any cows.

"Over HERE!" Susie hollers. "They're over HERE!!"

"OOOOKAAAAAY!" I yell back and drift in her direction, swinging my stick. I catch sight of a Holstein emerging from shadows on the other side of the pasture. Coming over a knoll, I see the rest of the herd, 30 or so cows. We never count. Nearly all of them are black and white, but there are three Ayrshires and two Jerseys. Susie is running at them and yelling "Get Up! Get Up! You Cows!!" Her voice is high-pitched and thin. They probably think she's a horse fly.

I fall in behind the herd, hassling stragglers, thumping their rumps with my stick to prod them on. My sister and I yelp to keep them moving. Now that they're rolling, they seem eager to reach the pond and I'm hoping one of them will fall in and swim. Their bones click continually while they walk – their hoofs, ankles, knees, hips – and occasionally one will cough or snort. We hear the first cows reach the water hole. They always sound a trumpeting bellow to the others, bragging about how good the water tastes and, "Ha! Ha! I may crap in it before you arrive!"

As Susie and I bring up the rear and look down on the water-baby cow circus from the pine-covered ridge, I'm startled by what I see at my feet. A disturbingly fat Water Snake is woven within the grass. Its dry scaly black-and-tan reptile body extends a couple feet from a clump of tall grass while its head looks out the other side. Wait a minute! That's no normal snake's head! It looks like a bug-eyed toad! I can't believe my eyes and don't dare step back. Nothing is moving. Not even the snake creature.

"Susie! Shush!!! Don't say anything," I whisper loudly, beckoning her to my side. "What does that look like?" I ask.

"NO!" she exclaims and jumps back. Leslie has arrived to deal with the cows. His harsh low voice rises from below. He does not care about a toad snake, but I tell Susie to go get him.

I still haven't moved and neither has the snake. I bend down to take a closer look. Its head is definitely like that of a toad. It blinks its bulging froggy eyes and twitches, first the head, then the tail – just a jiggle – and it stops. The creature becomes perfectly still. So I take my stick and push aside grass in the clump because I want to see how the head is hooked to the tail. Once again, the head wags back and forth and the tail twitches.

I poke the grass again and Oh-this-is-TERRIBLE! It's a snake swallowing the hind end of a toad and it's up around its stomach, headed for the front legs! The second set of eyes – the snake's real eyes – are blank black dots bracketing the toad's ribcage. The dusty serpent's head is deformed from being stuffed with toad. Its jaws seem disconnected at its thick throat. The toad kicks its back legs inside the snake's throat. The snake slowly shifts the full length of its body within the grass.

I'm focused with dazed fascination and don't know what to do. Seems like Susie ran to tell Leslie a long time ago. I've watched this snake and toad for so long that my neck and shoulders have begun aching. The whole place has gone silent. What happened to the cows? Gone. I knew Leslie wouldn't come up to look at a snake. He says: "I'm the hired man, not your numbskull babysitter. The hired MAN!" They have left me here alone and I'm getting tired of watching not much happening in this clump of grass.

I'm on the toad's side and take my stick and tap the snake's nose. The toad senses a shift in the action and wriggles again and wiggles forward, just a little. But this time the snake can't do its usual munch-munch move to regain some stomach because I'm holding its jaw back. So the toad gives another heave-ho and manages to pull its right rear leg out from the snake's throat. Being motivated by hope for survival, the toad begins kicking the snake in the eye. The snake's tail whips back and forth as its brain registers a loss of toadage. I hold steady with my stick, wondering if the snake has noticed my bare toes quivering nearby. The toad is obviously fatigued, but makes one last yank to slip its left leg from from the restrained chompers ... and quickly hops into another clump of grass.

Now, it's just me and the snake. I hold my breath while relaxing pressure on the stick. To my great relief, the black thing takes a turn to the right and silently disappears.

I try to get one last look at the toad, but I'd have to poke the grass clump to see it. I've done enough poking for one day.

Nothing neat like this ever happens when I bring in my Peterson grandparents' cows – "Grammy" and "Grampy's" cows in nearby Fremont, where my mother and Aunts Alberta and Ann and Uncle Philip grew up. In the first place, they'd never let me go into the pasture alone because they have a bull. And second, why would I want to be walking around with toad snakes when I could be riding behind Philip on his mare Sprite?

I like horses better than cows. Maybe I'll get one some day.

Part Three – Reduced To Rabbits

Kids grow up. Our son, Otis Lyman, graduated from The University of Maine as a mechanical engineer and married his high school sweetheart, Melanie Fike, on June 15, 2007 (officiated by Grandpa Ted Clapp). Mel had also graduated from U Maine and went on to become a Registered Nurse. Our first grandchild, Rachel Elizabeth, was born October 10, 2009.

Second son, Amos, came to New Zealand with us in 2003. He attended Southern Institute of Technology for a year, earned Welding Certificate. Then Uncles Albert and Matthew fixed him up with '87 Cadillac and he toured and worked around USA for five months.

Amos Theodore graduated from The University of New Hampshire in 2009, majors in History and Art.

'The Pump House' with Amos's Caddy.

The way we were ...

29. MY EARTH MOTHER

No secret. The 1960s were turbulent years, locally, nationally and globally. Kids today think their lives are "troubled." But I'm pretty sure it's a different kind of trouble than what we had in the Sixties. Back then, it felt like GOOD trouble. As a young adult, I was hopeful and energized. Events were upsetting, but not so sinister or complicated that we couldn't do something to make improvements. We spoke up and people seemed to listen.

For me, today's youth seem to live in an atmosphere where troubles are called 'issues' and these issues are not very exciting, just heavy, dull and depressing. Relieved and grateful for the way our own offspring are turning out, I'm past blaming anyone for being apathetic. But I do worry that many of 'Generations X' and 'Y' don't seem to know WHAT they believe in, and I'm not sure that's healthy. But what do I know?!

A young friend wrote to me recently: "Government = Crooks. With billionaires vying for control of our nation, do you really think any candidate is 'For the People'? Truth is, the fat cat politicians have bigger fish to fry than you or me and they're no longer afraid of the voice of the little guy. The 'angry mob' days are gone. You can't get Americans worked up over much of anything anymore. You can talk smack all you want about the administration and the FBI isn't going to come knocking on your door. They know that no one who counts is listening to what you have to say. We are frogs chirping in the night." And he signed it: "Ribbet. Bob"

Bob's words make me sad because there is so much that needs to be done to reverse the numbing effects of human nature: greed, hatred, lust for power and dominance over the environment, addiction to video images and things that will kill us.

Americans seem stupefied lately.

But it wasn't like this following my high school graduation in 1965. There was an electrifying sense of responsibility, a belief that if you got out there and kicked butt, things would change. It was dangerous back then, too.

Part Three – Reduced To Rabbits

In 1963, John F. Kennedy was assassinated in the middle of Dallas with the world watching on TV. Nearly everyone was sad about that – everyone I knew, except my Christian-conservative ceramics instructor who stunned me the day of the killing. "Someone should've shot the bastard sooner!" Mr. Jones said.

Then, the replacement president, Lyndon Johnson, intensified our assault on Vietnam, and nearly everyone was mad about that. "Rolling Thunder" air strikes continued for three tormented years as my male classmates were either drafted or enlisted in the military hoping for safe assignments. The Vietnam War and attacks on Cambodia continued until we had blown up 3-1/2-times as many explosives as were used during World War II, plus we tried to defoliate the whole place with weed killer.

This didn't win the United States many friends nor garner much respect.

On the home front, civilians died to show their disgust. It was something they could do. Alice Herz, an 82-year-old survivor of Nazi terror, set herself on fire in Detroit to protest the bombing, as did a Quaker and a Roman Catholic outside the Pentagon and United Nations, respectively. Later, National Guardsmen shot into a crowd of student protestors at Kent State; wounded 13; killed four. A confusing debate began over "patriotism." Who was more patriotic than whom?

Civil rights activists were murdered across our southern states as Martin Luther King Jr. led marches. When he called the United States "the greatest purveyor of violence in the world," he was assassinated.

And so was Robert Kennedy.

Despite our people dropping like flies, I feel fortunate to have been awake and participating in The Sixties.

Even on the farm where our spirits were generally dulled by the demands of animal husbandry, there was a sense of unrest and rebellion. Susie and I listened to JFK's plea to: "Ask not what your country can do for you. Ask what you can do for your country!" We decided to join the Peace Corps. Susie said it was her idea, but I was the one who joined and almost went to French West Africa.

Our parents held their breaths and hoped for the best. The farm was hanging on and they hoped their kids wouldn't go off the deep end. Mom was noticeably restless during the Sixties. She could hardly wait for me to get my driver's license so we could leave town.

"We've gotta get out there and see more of the world," Mom said. "We'll drive to Yellowstone Park and see what it's like. I want you kids to learn how to travel."

She was serious and I believed her.

I had been driving hay trucks and tractors ever since I could reach the pedals, but can't recall any personal longing to obtain a driver's license within hours of my sixteenth birthday.

It was Mom's thing.

She needed me legal to hit the road as soon as school let out in June, 1963. Her scheme was to take us westward to the nation's premier National Park via Osage, Iowa, where we have friends, Uncle John's family. She would stash 'The Babies' (Beth and Erica) on Mast Road with the Mingo family and leave Dad to milk the cows.

Mom made me practice driving our '55 Ford wagon up and down the road in front the house, turning through gateways on either end, wheeling the eight-seater around in the fields and signaling to nobody as I returned to our driveway. Parallel parking made no sense to me. I could back a two-ton rack body loaded with hay through the barn door using mirrors, but why would I ever have to back into an end-to-end parking space? Even Manchester was straight-in-and-out parking. Nevertheless, I practiced parallel parking in the field between foolish stacks of baled hay.

It snowed big time on the eve of my driver's exam. The roads were hardly plowed as Mom and I drove south to Plaistow. Then, we got a flat tire within a mile of the police station where licenses were given out. So we were late. But I whipped through the written part of the exam and passed that, so qualified to take the driving test.

Well! I did okay through town using my signals, changing smoothly through the gears and all. But then the examiner had me drive out a side street with snow plowed high on both sides. He told me to turn around in a gateway and head back.

I thought: "You've gotta be kidding!" The gate wasn't open and the packed snow provided no margin, plus the station wagon was long and lacked power steering. You know where this is heading. I went back and forth no fewer than seven times, nervously, before finally shifting the barge. The examiner was not amused. My arms were too tired to parallel park in front of the police station.

I failed.

"You didn't even turn the wheel!" Mom said, disappointed and, no doubt, suspecting that Susie could've done better.

It took all winter to get my license. Three tries. So what! I learned to drive somewhere on Route 80 in upper New York State when six of us went cruising with Mom: Susie, Matthew, Phyllis, Albert, Cindy and me.

The trip west changed my life.

I loved the routine of stopping at rest areas for lunch. Ice cubes must not have been invented yet because we never had any in our gasketed Coleman cooler. Our sandwich ingredients were contained, but assumed car temperature. We stuffed Wonder Bread with baloney garnished with sliced Velveeta cheese spread with Miracle Whip and mustard. We refilled our 1.5-gallon plastic Thermos with tap water colored with Kool-Aid, sweetened with a cup of white sugar.

It was great!

Toward night, we checked our road map for tent symbols where we could camp. Once, we settled along the edge of a farmer's flooded potato field. We were the only campers. There was so little space along the fence that we could hardly avoid the rutted muddy access road. Complaining and relocating were the furthest things from Mom's mind. She said we should feel sorry for the farmer having it so wet. So we pitied the poor farmer.

We slept on blowup mattresses in a different place every night. We camped on the Canadian side of Niagara Falls, then dropped down to proceed across the Midwest and through the Badlands and Black Hills with Mount Rushmore, which I've returned to three times.

Another time, we tried to camp with a traveling circus in Chicago. Mom was all for it. She said we'd blend with the gypsies. But as we blended and began erecting the tent, it became apparent that mobs of children had been let loose to preview the animals and freaks.

"It's not worth it!" Mom declared. "Let's move while we still can."

I remember waiting an hour for a tour to begin at a Waterloo, Iowa slaughter house. Mom thought it would be good for us to see a modern meatpacking facility. We arrived between tours. Outside the industrial abattoir, the parking lot was oppresively hot and humid and smelled terrible. We tried to be brave good sports, waiting as we had told the office staff we would.

But boy! Did it ever stink!! We wondered how the cows would be killed, with a pistol, rifle, stun gun or we didn't know what. I didn't feel well and was probably coming down with The Travel Bug.

I can't recall which of us said: "I don't want to do this." But it was all Mom needed to lead a hasty retreat into the TravelAll. We were on the road again without even telling the office.

The adventure of abandoning undesirable camping spots and evading creepy situations balanced the stability and resourcefulness we felt from efficiently setting up and disassembling camp gear.

Later in the trip, we returned to Canada and camped at Lake Louise in honor of Mom. We loved Yellowstone Park with its continental divide and wildlife, the magnificent Rockies, the geysers and mineral pools. We went through the Grand Tetons and Jackson Hole.

The most exciting thing that happened in Yellowstone was when Matthew and Albert disappeared as soon as we began setting up camp. They promptly returned, each smiling and dangling a trout from their fishing lines.

Mom had no more than pointed to the 'No Fishing' sign beside the road when two cars crashed into each other, RIGHT THERE beside us. As park rangers convened and we kids got nervous, Mom said: "Quick! Let's fry 'em up!!" The evidence was promptly dispatched to our stomachs.

A week camping in Yellowstone's hydrothermal basin changed my outlook and indeed established a few habits for life, like keeping a journal and taking lots of pictures. The writing and photography went hand-in-hand for noting details and cropping nonessentials. To save film, I made a rule that a person had to be in each picture. I consciously focused more on people than scenery. I looked past finishing high school and decided to spend time in each of the 50

United States. I knew I could buy a car and keep its wheels rolling.

Going to and back from Wyoming, we spent time with Uncle John's family in Osage, Iowa. One of his nephews, about my age, Mike Johanns, grew up to be Secretary of Agriculture under George W. Bush. He even came to the farm in Epping! Pretty funny, huh?! Now he is a U.S. Senator from Nebraska. Republican, of course.

Sometime after our summer-long trip to Yellowstone, Mom befriended 'Jimmy and Jo', a young couple from the middle of New York City. They both had really strong nasal accents and long dark hair. Jimmy's was longer than Jo's. He wore it in a pony tail. Mom met them at a weekend conference in nearby Durham at the University of New Hampshire. She had gone to learn about organic gardening. Jimmy and Jo exhibited a tipi they had built and explained themselves to whoever stopped by to listen.

Mom was a good listener and liked what she heard about the New Yorkers' need to escape the skyscrapers and pavement, how they intended to reconnect their roots to Mother Earth by studying and trying to live like Native Americans. Mom invited them to set up their tipi in one of our secluded back pastures. Still called "The Tipi Field" after 40 years, it now belongs to Phyllis as part of 'Edgerlys'. Mom offered to show the New Yorkers how to bottle food for the winter if they helped her in the garden and taught her whatever they learned about Indians.

So, that was "different" and gave us all something to think about as we sat on blankets while sipping herbal tea in Jimmy and Jo's white tipi. Later, Liz, a friend of theirs, also set up a tipi out there.

During the Sixties, my parents assumed two stances while demonstrating their commitment to preserve the agricultural integrity of our home – defensive and attack.

Mom showed defensive strength in always holding the fort to model a farmer's unapologetic work ethic, demonstrating everything from sewing and gardening to carpentry and competing in spelling bees. She showed us how to volunteer at church and school. We were taught to say: "I'll be glad to!"

And Dad poised for attack, never hesitating to "wade in amongst 'em" to defend our life, liberty and pursuit of happiness within "these Great United States of Amerikey." He stood ready to jab back and defend the farm because he, more than any of the rest of us, sensed our vulnerable situation. Dad saw dark economic and societal forces lurking along the wooded roadsides, ready to pounce and swallow us. He said bad times were coming to Epping because there was no way to stop people from acting like sheep and wanting progress.

Recently, Dad said our little town went sour in the early 1960s when the selectmen decided there should be building permits. He said that's when citizens stopped smiling inside the Town Hall because "that's where many gooses have been cooked."

I have to agree.

Today, Epping seems like most other towns where it is a challenge to get people to run for public office. People don't seem interested in volunteering for local government. Everyone is too busy. But back in the 1960s and '70s, Epping elections were hotly contested with several candidates running for each opening – selectmen, budget committee, school board, water and sewer commission, trustee of library, trustee of cemeteries, overseer of the poor. I could be wrong, but it seemed that even chiefs of police and fire department were elected positions. Politics were hottest in March just before Town Meeting.

Political flyers clogged mailboxes; 'Vote-for-Me!' signs were posted at road intersections. Some were always stolen. Letters to the editor made for good reading. Epping elections always boosted sales of 'The Exeter News-Letter', since Epping news was more exciting than Exeter's.

With the advantage of hindsight, our small-town politics were SO petty, considering what was happening with the rest of the world. General unrest was sweeping the United States and there we were in Epping drawing cartoons of each other over zoning ordinances.

Racial integration was the first social problem to enter my consciousness when President Eisenhower sent 1,000 soldiers to Little Rock in 1957. The image in my head was Arkansas state police standing in the way of nine black kids with National Guardsmen holding the school door. There were riots in Georgia, Mississippi and Alabama with much talk about a state's right to discriminate if a majority of their voters wanted to.

To Epping's credit, we students studied "civil rights" and stood up with our weekly "current events" reports on Freedom Riders and 'Negro' voter registration literacy tests. Each year, through junior and high school, there was a "Civics" class to discuss what 'The South' was going through to integrate schools and housing.

James Meredith was a headline when President Kennedy sent U.S. marshals to protect him while attending the University of Mississippi (in 1962). Then the Ku Klux Klan bombed a Baptist church and killed four little girls in Alabama, and (in 1964) three civil rights workers were murdered in Mississippi (one black plasterer's apprentice and two Jewish students from New York City). People were being beaten, shot, bitten by dogs and thrown into jail where nobody denied harsh treatment.

When I could, after a year at New Hampshire College, I transferred to a school in the integration action zone: Texarkana College on the Texas-Arkansas state line close to Oklahoma and Louisiana. I wanted to see what was going on. As a Work-Study student in the registrar's office (1966-67), I was privy to continual inappropriate comments by staff about the black students.

One day, the assistant registrar, Ruby Nell Hatch, said to me: "Y'all must think we're awful!" And I thought to myself: "Well. No worse than the girls in my freshly-integrated dorm!" I kept my eyes open and my mouth shut.

These were not America's finest years. And, of course, we had the Vietnam War and mounting national debt, amid a lack of confidence in the government.

In 1968, the U.S. presidential election was joined by the notorious Alabama Governor George Wallace, who was against "big government" and racial integration. Wallace came to New Hampshire FOUR times for the nation's first primary. He vilified blacks, students, and people who called for an end to the war in Vietnam. He shouted things like: "I draw the line in the dust and toss the gauntlet before the feet of tyranny to say: Segregation now, segregation tomorrow, segregation forever!"

If it hadn't been for George Wallace siphoning votes as a third party, we might have elected Democratic candidate

Hubert Humphrey as president, instead of Nixon who won by less than a percentage point.

Nixon was immediately unpopular as he escalated the war. (Half of American Vietnam War 58 thousand fatalities occurred during Nixon's watch.) For some reason, I was drawn to those who spoke out against what I saw as wrong – especially the killing and defoliation in Vietnam. As a UNH senior, I joined anti-Vietnam War marches in Concord and Washington D.C.

Sad to say: Dad voted for Wallace. His was one of the siphoned votes because he thought it was funny that Wallace called federal bureaucrats "pointy-headed intellectuals"! And now Dad has voted for John McCain in the primary, a Republican, rather than support Barack Obama.

It's not that Dad is a racist.

He's just a BRAT!

Little wonder I felt like misbehaving in the Sixties! I still don't regret joining demonstrators to protest the war, segregation, and the tremendous conformity of the Fifties. I'd be disappointed in myself if I didn't.

Uncle Eddie and Aunt Ann fretted for fear I had become a communist. Or a hippie, certainly.

But Mom? Well. Jimmy and Jo weren't the only ones to tell me: "Your mother is a better hippie than YOU!" (Referring to her Earth Motherliness, I'm sure.)

Dad running sawmill for farm building materials. Not fitted for safety, but produced plenty of square timber. Once, a tooth flew off the large circular blade (spinning lower right corner) and cut a skid mark across Dad's forehead as it flung and nailed his cap to the wall.

Historically speaking ...

30. N.H. DAIRY INDUSTRY - PRODUCTION

The Harveys participated in nearly every phase of America's developing dairy industry – up until agri-business and factory farms took over.

The Civil War (1861-1865) prompted a shift from rural to city living while limited refrigeration and transportation prevented distant farmers from shipping milk to built-up populations. So, in New England small family farms were the norm until railroads opened urban markets to rural suppliers.

Around the turn of the century, 115,036 cows were kept on 22,519 New Hampshire farms.

I'm surprised to learn that one of my 4-H project calves came from a famous herd. 'Bonnie', a registered Guernsey, came from the Hood Farm in Derry, New Hampshire. She was my 'show' animal, the one I trained to walk slowly and pose for judges at fairs.

Apparently, in 1846, Harvey P. Hood left his family's farm in Vermont to establish a dairy closer to the Boston market. He collected fresh milk from farms around Derry and traveled south with it each day into Boston. In 1860, Hood arranged to have a special train car built that held 8,000 quarts of milk to be chilled on ice during the summer and kept from freezing with a wood-burning stove during winter. (*Selitzer, Ralph: The Dairy Industry in America, 1976, p. 113*)

But Mr. Hood couldn't supply it all.

On into the Depression, most urban populations were subjected to poor-quality milk furnished by often unscrupulous in-town distributors who paid little attention to sanitation and tampered with the supply. Fortunately, limited technology and logistics spared Epping farmers and most other small-town families from food quality problems experienced in cities. The milk supply was kept close to home. Country people continued to grow their own food and buy locally. My great- and grandparents' cash crop was apples.

Until I was born in 1947, the Harveys kept just a few cows for home supply. They had no concern for the developing Boston milk market. As a child, I remember both

Dad and Grampy Peterson milking cows by hand. Laughing, Dad squirted it directly into a cat's mouth as it sat behind the cows.

Both families, the Harveys and Petersons, owned cars and trucks. Even then, each farm had a tractor. The Petersons hauled blocks of pond ice from Massabesic Lake near Manchester, while the Harveys used an electric compressor to cool 10-gallon jugs in a water bath.

Though a Boston-bound train passed through Epping daily, I don't remember our milk leaving the farm other than in Clover Leaf Dairy's enclosed truck. The bottling plant was just 20 miles away with a collection and delivery territory straddling the state line between New Hampshire and Massachusetts. Residents received fresh bottled milk on their doorsteps.

At some point during the late 1950s, Clover Leaf shut down because home-delivery had declined and it was too expensive to upgrade for bulk handling.

By then, nearly all milk was pasteurized and homogenized.

Clover Leaf processed milk, but I was not impressed. I remember refusing to drink school milk because it tasted "cooked." Teachers said it was good that it tasted cooked because all of the germs had been killed. What did I care about germs? I had been eating them all my life and knew they made the milk taste better.

During the 1950s and 1960s, refrigerated bulk tanker trucks began hauling milk to Boston from upper New York State. Competition with Northeast suppliers guaranteed that farmers would receive the lowest possible price if the Department of Agriculture didn't step in to control the market.

Many dairy farms went out of business because the list of necessary technology upgrades was long and daunting. With the price paid by milk companies stagnating at about $10 per hundred weight, farmers saw little hope for return on their investment.

So they quit.

My parents optimistically considered and accomplished each recommended upgrade. Here is what I recall:

'OLD BARN' IMPROVEMENTS: 1) Poured concrete for first milk room, 2) One water-bath refrigeration tank, 3)

Reconfigured Old Barn tie-up to provide more space as heifers calved ('freshened' and began milking).

'NEW BARN' IMPROVEMENTS: 1) Built modern tie up, but we still had to carry milk to the 2) New milk room which had two water bath tanks for 10-gallon jugs, 3) Concrete floors, 4) Drinking cups, 4) Convenient baled-hay storage overhead, 5) Two gutter cleaners, 6) Modern Surge milking machines, 7) Overhead bulk grain bin for pelleted feed hauled in from Midwest, 8) Another new milk room to accommodate, 9) Bulk milk tank that we filled with buckets from where cows were milked in tie-up (other farmers installed dumping stations with hoses to transfer milk from tie-up to bulk tank).

'OLD BARN' & 'NEW BARN' MERGED: 1) Old Barn reconfigured to extend two rows of "tie-ups" from New Barn, 2) Trench silo built to store fermented corn and grass silage handy to 3) 70-foot-long feed alley, 4) Always pouring more concrete to keep herd out of mud.

MILKING PARLOR: 1) New Zealand-style Herringbone, two rows of six cows, 2) Overhead pelleted grain storage to supply parlor hoppers, 3) Holding area.

POLE BARN & COVERED COW YARD: 1) Reconfigured Old Barn again to merge with Pole Barn for 80 cows, 2) Cow yard where herd was confined, 3) Ever-expanding roof over the cow yard, 4) More comfort stalls, 5) Automatic water stations, 6) Programmed supplemental feed doors for high producers.

IMPROVED MANAGEMENT: 1) Artificial insemination (Mom said it was too dangerous to have bulls around children), 2) Herd production records with monthly inspectors, 3) Feed programs, 4) Pesticide, herbicide, fertilizer programs, 5) Crop strategies using hybrid seed and chemical fertilizer, 6) Finance management schemes, 7) Herd health and vaccination programs, 8) Always larger, never new tractors and crop equipment.

And on it went. I've probably left out half the improvements.

There's more to farming than one might expect. We tried to do it all. And, in the end, we weren't even contenders!

N.Z. Southland 'Fonterra' milk trucks are tandem tankers dispatched from processing plant in Edendale where the Prime Minister recently opened the world's largest milk drier. Nearly all fresh milk is immediately powdered for dairy product export. Fonterra runs 60 of these trucks around the clock in just this district. Below: 'My' cows in Riverton herd of 400. Herringbone shed, 28 on a side, enhanced with molasses troughs to give the herd more energy and brighter outlook on life.

Part Three – Reduced To Rabbits

Our life-long friends and neighbors
Robert and Amogene Kimball of Old Hedding Road.

Sisters Ann and Louise in Hawaii, 2009

Uncles, aunts and parents: Philip & Virginia (Peterson), Alberta & Roland (Schreiber), Ann (Burtt), Louise, Dan
a week before Philip died.

Time in the kitchen. Grammy and Christa made pies. Susie brewed coffee.

Sunrise as seen from deck of Dan and Louise's new house.

Steve Martel and Erica Harvey were married on top of hill June 7, 1998. Christa Weaver was their flower girl.

PART FOUR

To Protect The Innocent

The way we are ...

31. PHYLLIS'S WILL AND TESTIMONIAL

The Sisters' Refrain ... (please)

We all could die any day.
And land could get snatched away.
So make out your will
'Cause they're poised for the kill.
But you keep your kidneys,
Okay?

Susie made an appointment with the farm's LLC (Limited Liability Company) accountant and told Phyllis to go make out her 'Last Will and Testament'.

Although the accountant's wife handles our legal affairs, it is her husband (the accountant) who shows us where to sign tax documents in December. Neither money nor information is exchanged. He simply flips pages and points where to sign by yellow arrow tabs, about six times.

This third-person arrangement is a company policy. While making out my own will, I asked my lawyer to clarify why the farm's lawyer hasn't explained the LLC to we share holders. It was determined that our parents don't think annual meetings are necessary. No. I cannot transfer half my shares to Amos as a graduation present – and, yes, the folks are comfortable being represented by Susie. Indeed, seven years ago, the lawyer sent us each a letter saying she will NOT talk with anyone other than Susie or our parents about Applehurst Farm business.

So, that's why when Susie tells us to go, it is with trepidation that we sign our names for the accountant.

And, yes. It was trepidation that Phyllis felt when Susie told her to get her butt down to the law firm and put her affairs into order.

"Why did you even go?" I asked my 57-year-old sister, amazed by her recounting of the episode. She said Susie was probably right, that we COULD die any day and it would've been trouble if she failed to keep the appointment.

"Besides," Phyllis said: "It has always been this way. Susie has always bossed us around. Don't you remember how it was with your chickens?"

No. I guessed I didn't remember. The forgetting thing, you know.

Phyllis said that, back when we were small, I "went off somewhere" and gave my chickens to Susie. Phyllis said Susie made her and Matthew tend the flock, morning and night. "We fed them and collected, washed and boxed the eggs. Then SHE sold the eggs and kept the money."

"It's just the way it is," she continued. "And don't you remember how she used to sit on us in front of the house, on the crushed stone, and pull our hair and sing: 'Cry, cry! See if I care! Go tell Mommy and I STILL don't care!!'?? Phyllis chuckled, alleging that Susie's role model was Barbara Stanwyck of TV show 'The Big Valley' and 'The Thorn Birds,' calling her the 'big boss woman of the lands!'

When I told her I didn't remember any of that, she said I must've missed my training.

Best I can tell, we Harvey kids all tried to train each other (and some of us haven't given up yet). How could I respond to Phyllis's accusations? I probably sat on her myself! And maybe the rest of them, too.

I was no prize.

Susie used to wag a crooked finger at me claiming I had twisted it during a fight. "Just look what you did! I'm deformed for life!" she scolded.

But I haven't sat on anyone lately. And, trained or not, damned if I'd be showing up at a stand-offish lawyer's office just because Susie told me to!

But, apparently, it's different for Phyllis. Having already agreed to be evicted, why WOULDN'T she do something as simple as report to the attorney's office to make out her will?

So. Phyllis met with the accountant, but ended up running out the door halfway through the appointment.

It was the first thing Phyllis told me when we returned to Epping last summer. She said Susie and Erica were mad at her for THAT, too.

"They went to 'all the bother' of setting up the meeting and here I am 'ungrateful enough' to just waste the offer of a 'Last Will and Testimonial'!" Phyllis said.

I told her it's a 'testament', not a 'testimonial'. But my sister often coins words for effect and reckoned (to her) it felt more like a testimonial.

Phyllis said the accountant expected her to designate heirs and say whether or not she wanted to go onto life support and whether or not Tufts Medical School should have her kidneys. And he told her to hurry up because he didn't have all day. He suggested that her land should be returned to the LLC, presumably, AFTER she dies.

"I told him NO. I wasn't ready to be giving away all my possessions and I don't even know who to give them to. I asked him who Susie was going to give hers to, since neither of us have children. He wouldn't say."

Phyllis said a wave of panic passed over her. She abruptly stood and left the building. Now she's in trouble because Susie said it was 'just business'.

"I try to get along," Phyllis said. "But I couldn't go through with it."

Phyllis June Harvey

'Edgerly Place' lawn ornament.

The way we were ...

32. IF YOU HAD A MILLION DOLLARS

Nobody wants to see the farm dissolved or sold for development. Yet, we can't agree on a strategy for its survival. Conservation groups have approached us urging enrollment in easement programs. But it's not as if we need an income tax write off and our parents aren't inclined to sign anything away.

Estate planning is complicated and becomes emotional as we consider paying a several million-dollars-worth of imminent federal and state inheritance taxes. It's not a question of whether or not we will be able to keep the land. We have to decide WHICH land can be sold quick enough to pay off the Internal Revenue Service.

A family joke lies in the wisdom of brother Albert's fourth-grade essay. The topic was: "If you had a million dollars, what would you do with it?" Albert wrote that he would use the money to pay the Epping property taxes. His teacher said he seemed 'troubled' about it.

We were all troubled, especially while going further and further into debt to maintain a modern dairy operation that grossed $100,000 a year from bulk milk sales, but was dragged down by an annual tax bill of $30,000 ... ('back when money was WORTH something!' as they say).

We might've done better if we had taken advantage of 'current use' tax legislation that the State of New Hampshire adopted in 1968. But Dad didn't believe in it and refused to sign up. He wouldn't listen as we told him we should take advantage of the deal, since we didn't plan to develop anything anyway.

He said it was against his 'principles'.

I'm not sure what he expected the Town to do when he wouldn't accept the tax break, but I do know what they did. They gave other compliant landowners the discount and increased taxes on everyone else because that's the way it works in New Hampshire. Each town has to cover its own expenses. The Granite State (still) has no sales or income tax. No chunk of broad-base tax revenue comes floating down from the capital in Concord. So: "The money has to come from SOMEWHERE."

Thanks to current use taxation, we had the highest tax bill in Epping. And we couldn't afford it. I hear that – to this day – Dad's friend Bob Bonser's heirs still refuse to sign for current-use taxation, continuing to call it an "unConstitutional communist plot," plus the Town of Nottingham still holds liens on their property and accrues interest on penalties charged in the 1980s.

We weren't THAT stubborn. Dad held out for just a couple years.

Still, back then, Dad kept borrowing money to cover our operating expenses and (seemingly irrationally) continued buying land until we had over 2,000 acres.

He was forever hassling the two Exeter banks (Exeter Banking and Rockingham National), playing one off the other, opening and closing checking accounts and securing short-term notes to cover his 'vulcanized' checks. The government's Production Credit Service was the preferred lender, but we were maxed out there and begrudged them their monthly draft off the milk check. They were not inclined to support Dad's land-buying habit.

Our credit rating was lackluster because our part of the delinquent taxpayers list took up a whole page in the 'Epping Annual Report'.

We were land rich, but always felt poor.

And taken advantage of.

A classmate called me: "Poor Little Rich Girl." It stung my ego, but, really, how far off WAS he? Seems we Harveys share a deep terror of poverty coupled with equally deep shame for not being poorer than we are.

1994: Red Oak Hill 'Cornball II': Hayride wagon and inflating hot-air balloon (probably 'Nasty Habits' from Raymond). The McPhee Family now hosts an annual music festival at their farm. Their 'High Note' scholarship fund-raiser attracts many balloons such as this one. Our two Cornballs were free and benefitted nobody, just cheered us up.

The way we were ...

33. TRANSIENTS WE'RE NOT

Two years after launching the "Cindy-Lynne" the Schofield family wasn't in town anymore and Dad had nearly died trying to visit them at their new home in Scituate, Massachusetts.

It was the story of our neighbors' lives: They came ... and then they moved away.

Granted, we Harveys aren't what you'd call 'transient'. We wouldn't leave home if it was on fire. So how COULD we grasp what it would be like to 'just up and move away'?

These new people would arrive and fold comfortably into our routine, stop by the farm almost every day, and give us Christmas presents. Babies were named after each other's kids. We'd invite each other over for cookouts. There was an annual 'Neighborhood Picnic' attended by people who lived even further away than Aunt Ann and Uncle Eddie's (on the back of the hill).

Families with kids our age carpooled to Hampton Beach. One summer, we went to Lake Winnepasaukee to water ski with the Sullivans. Sometimes, we met friends at Adams Point for outings. Our neighbors did normal things like direct the church choir or run for School Board or 'Selectman', or something less controversial, like Supervisor of the Checklist or Overseer of The Poor. They'd prove themselves to be loyal friends and capable upstanding citizens.

And then, in what seemed an instant, they MOVED! GONE!! What would a Harvey know of such carry on?

We were saddened and grieved whenever we heard that a neighbor intended to leave. Probably that's why the telling was such a big deal and Mom was usually the first to know. Or maybe we just THOUGHT she was the first to know because she'd say, "'So And So' came and told me NOT TO SAY ANYTHING. But they're moving." Then she casually swore us to secrecy: "Now, don't YOU say anything!"

We began eyeing our friends with anticipation, waiting for them to tell us directly.

I remember standing in a tangle of blackberry bushes with Cindy Schofield wondering if she even knew this was her last summer. The sun was hot on our backs and radiated off a large riveted steam boiler in the junkyard behind our shop. Berries grew especially well next to the heat sink of rusty scrap metal. Bees and hornets worked the bushes with us. Their nests were inside the tank.

Cindy and I were about eight years old. She was one of my four best friends: Mary, Sherry, Cindy and Frances. I couldn't mention her leaving town if she didn't tell me first. And she didn't, so I tried to act as if nothing was wrong. We had been shedding blood over those berries for a week, scoring scratches while extricating ripe fruit from among the wretched stocks – wild dark-staining berries, smaller than what is available today, but more flavorful.

Cindy Schofield and I were in the same grade in school. She knew the horrors of first grade with Mrs. Beecher. She was there when Larry Emerson had his mouth taped shut and sat in the supply closet with the door closed, and when Linda Levesque was thrown under the sink just as another teacher came in, and Mrs. Beecher said: "Oh, My Dear! How did you ever fall down like that?!"

Though our eyes were supposed to be shut, Cindy and I peeked at each other during Mrs. Mead's morning 'Lord's Prayer'. We thought it was so funny the way Mrs. Mead pushed us on to say "...For Thine is the kingdom, and the power, and the glory, Forever and Ever and Ever and Ever. Amen" to mock the Catholic kids who had to end it with being "delivered from Evil."

Cindy and I had learned how to behave together. She knew where my secret hiding place was in the old barn, behind the sawdust pile at the end of a tunnel that dumped hayseed onto our heads and down our shorts. It was Cindy who was on the front when our toboggan ran into the barbed wire fence by the pond. Good thing she wrapped a scarf around her neck three times or her head might've been cut off.

I remember feeling 'homesick' for Cindy there in the junkyard berries. And she wasn't even gone yet! Tears blurred my vision and a slippery little snot dropped onto my arm as I remembered how Cindy bit Lisa Young on the shoulder when she wouldn't move over on the bus. And the way Cindy didn't cry too much when she was riding on the back of Mary's bike and her foot got stuck in the spokes.

It was probably the same for my parents, similarly losing their best friends. Parker and Nancy Schofield were just the first in a series of occupants of a former Harvey house that became the 'Haddock House' when Mrs. Haddock bought it. Dad says they used to call the widowed woman 'Mrs. Fish'. I don't know what the fish lady did for work, but subsequent breadwinners were mostly airline pilots who commuted to Logan in Boston.

'EPPING' must've been posted in the pilots' airport lounge as a desirable place to live, since nearly a dozen airline families bought homes in town. One Haddock House couple – just two people in a very large house, perhaps a 1960s premonition of 21st century 'McMansions' – was a pilot and a 'stewardess' who bought two of everything the whole time they were here. He'd buy a new car; she'd buy a new car. He'd get a snowmobile; so did she. A Lay-Z Boy recliner; she'd get one for a lazy girl. They said they did it so if they ever divorced, it wouldn't be hard to divide the goods. He ran for Selectman, but lost. She didn't bother running for anything and probably didn't vote for him. They soon divorced, packed their toys, and left town.

Another favorite Haddock House family was supported by General Electric. Or, was it Westinghouse? I forget. But, in 1959, Fritz Welhoffer was a high-profile electrical engineer dispatched to Portsmouth Naval Shipyard by Admiral Rickover to work on the 'Nautilus' and other dry-docked nuclear submarines. The Welhoffers stayed just three years but had an enduring impact.

We named my youngest sister 'Erica' after Erica Welhoffer; and they named their next-born 'Susan' after my sister Susie. It was the same jolly reasoning behind naming my sister 'Cindy' after Cindy Schofield. And 'Beth' (Martha Beth) after Epping's foremost 4-H leader Beth Bourne. And Phyllis's middle name, 'June', is for our babysitter June Allen (whose more beautiful sister, 'Mae', became Miss New Hampshire and had her autographed bathing suit picture framed over the meat counter at Epping Trading Center. (For talent, she milked a cow)).

Perhaps Mom was just running out of names as our numbers increased to six, seven and eight.

Bottom line: Our Haddock House neighbors all moved away. As did the retired railroad engineer and his wife, and two women who operated a catalog business from the garage until their computer crashed and they lost all the orders.

Seems a shame that so many people left such a handsome well-sited house, so grand and genuine, unlike today's mansion wannabes who elbow us from all directions. Built by settlers, the old Colonial is twice as big as our house. According to Dad, the Haddock House carpenters claimed its timber frame was so strong that "it could be rolled end-for-end across the field and not fall apart." It has 'Indian shutters' and wide pumpkin-colored floor boards honestly worn thin through the doorways. The ceilings are irregular, hand plastered, not mirror smooth. Its five brick fireplaces were obviously built to heat the rooms. The deep six-foot-wide opening and wrought iron crane of the kitchen fireplace is naturally complemented by a large beehive oven.

(It's not like me to indulge in this house envy!)

The Haddock House cellar is so big that you can't see from one side to the other. Massive masonry on the upper floors is supported by lower-level arched barrel vaults that make perfect brick cells for Halloween spooks. Long before haunted houses became popular, the Schofields and Welhoffers scared us with elaborate cellar ghost experiences.

Today, 2008, the jury is still out on our hundred or so new subdivision neighbors. We don't know what to make of them. Realizing that not even our most magnificent properties can hold their occupants for long, we fully expect these McMansion people to leave soon. But it's hard to say, since so few have stopped by to introduce themselves. Mom used to drag us all along to deliver welcome-to-the-neighborhood gifts to new residents – usually a fancy Bundt cake with a bouquet of violets in the middle. But now there are too many to keep track of.

"I can't be bothered!" she says in frustration. "Aren't Peter and Kim and Ingrid enough?" she adds, referring to two families she recently welcomed to the fold. Mom holds out hope for her vegetable stand.

"There's money out there!" she exclaims, fully expecting all the new people to yearn for her farm-fresh produce.

"When they see how bad the Wal-Mart stuff is, they'll come here," she says while stocking baskets and trays of tomatoes, potatoes, beans, broccoli, onions and squash. And, yes, her vegetables seem to disappear by day's end. So that's good.

Michael Coulp and Phyllis

Phyllis learned how to trap foxes from Malcolm Locke of Barnstead, who held the world record for awhile.

Mom decorating vegetable stand with 'corn shock', helped by Lawrence Lyford, Susie and Otis and Amos.

Phyllis and Michael Coulp with team of 'Hatty' and 'Otis' (Named after our son. Phyllis said it was a better name for a horse than a baby.)

Part Four – To Protect The Innocent

The way we were …

34. SLIPPERY SHEILAS

Not all neighbors worked out. There were a few slips. Harvey land sometimes got away and never came back, though some of it did. There was a man from Michigan who married the widow of one of our neighbors. He cry-babied and begged Dad "for the sake of Sylvie's peace of mind" to sell him ten acres with frontage closest to their retired farm.

"Sylvie can't sleep at night for fear of what you might do in that field," he said. So my parents agreed to sell it to them for the pittance offered by the man from away. And, sure enough! Within a year he had submitted an application to the Planning Board, asking to build two apartment houses on the property.

As a fledgling indignant citizen, still too young to vote, I wrote a letter to the board explaining what had happened, ending with: "I believe Mr. Arrogant's newly-acquired land is very ledgy, too close to open water, and generally unsuitable to sustain a lot of people wasting upon it. Therefore, I respectfully request that you watch him closely."

Uncle Eddie had warned my parents against the deal, saying this newcomer was a 'big talker' and couldn't be trusted.

But no, they wanted to be 'neighborly' and THAT acreage would not be coming back! Later, the man became a selectman and at one meeting announced that Dad was "so crooked they'll have to screw him into the ground when he dies" ... and he's the same "knowin' puke" (Dad's term) who blustered his way onto the school board and crowed about how he chose the cheap "shoe box" design for the new high school. He said it was "good enough" for us.

So much for being 'neighborly'! I figured he acted that way because he felt guilty over the land deal.

Then there was the affable young drug dealer from Chicago who shoehorned his way into our lives by offering Dad way more money than anyone could refuse for one of the three original Harvey home places – just over the line, in Nottingham, next to where our barn was torched. At the time, we needed tax money badly (per usual) and we didn't question its delivery in brown paper bags stuffed with small bills. Fortunately, all's well that ends well! As the genial gentleman was sent to prison with his big ego and Midwestern accent, he quickly sold the homestead back to my parents (minus the thousands-of-dollars-worth of drugs that state police dug up from under the lilac bush).

Score one retrieved yupp-ified 18th century dwelling! Susie somehow got her name on the deed, instead of our parents, probably because she was the guy's friend. Thus continued her ascension.

Then, also because we needed tax money, Dad agreed to sell 50 acres to a local developer. Actually, this has happened twice.

The first time – when it didn't work out very well in Lee – was when he sold to a carpenter who intended to build 'nice' houses and pay us off as he completed construction. Somehow, early in the process, the housing market went poopy and the carpenter couldn't make his payments. Plus his wife was divorcing him. So, no problem! He and Dad sat in the kitchen one afternoon "commiserating" and pondering "the sad state of affairs." And Dad, being the good guy that he is, forgave the man his interest payments and half the purchase price, to boot.

The settlement included a Kubota tractor, which Susie keeps it at her house. The builder finished whatever he was working on and resold our lots to another speculator. I still haven't driven by to see what it looks like. But I've heard that the resale price was so high that 'the Harvey land' raised Lee tax assessments on all other property along that road, the false assumption being that we had made a bundle.

But the second sell-off turned out better. My parents sold 50 acres on the back side of the hill to a developer who gave Susie and Phyllis jobs during construction. Phyllis had fun running heavy equipment and Susie got a new pickup truck with 'Susie' written in script on the door. I'm not sure what she did for work.

Still, here's the best example of 1960s land sales. Another Harvey homestead is a charming 100-acre (now a) horse farm at the end of a private dead-end road not far from where we live – the same place where I was too scared to visit Aunt Ann because of the closed gate. My parents sold the property to a chiropractor. The LaBarre family was nice and we became friendly growing up with their kids, Laura and Greg. We harvested hay for their horses.

Doctor LaBarre was a genuine warm-hearted man, a good neighbor whose hobby was rehabilitating damaged horses. Every Sunday, Epping residents sighed wistfully as the good doctor drove a buggy three miles into town to pick up his newspaper. Such a charming sight in a country setting: "Doc" with his black hat, black buggy and flatulent horse.

"A farting horse is a healthy horse!" laughed our friend during a regular stop at the vegetable stand. His jocular outbursts were loud and brassy. You couldn't help laughing along whenever Doctor LaBarre burst out.

But our neighbor wasn't laughing at the end. Doc had extremely bad luck with women. If he had anticipated how unpleasant things could become, he probably would've tried harder to stay married to his first wife (the mother of our friends) because the second wife, a real 'looker', promptly died of cancer. And then he married a woman who threatened to stab him with scissors while he slept and chased him around with a baseball bat while he was awake. Then, when she couldn't catch him (because he had come down to hide with us at the farm) she smashed windows out of his car.

We wouldn't have had much to do with her if she didn't start declaring herself to be a 'Harvey'. She actually told people – including me – that she was one of the 'Harvey girls' because someone had suggested that she looks like us.

But whatta ya gonna do?! Mom continued letting Doc sleep on the couch when he needed to.

Another of his third wife's delusions was thinking she was a country-western singer. A friend of ours who owns a recording studio assures me that the woman lacks talent.

This vocal shortfall was brutally apparent as she sang "The Funeral Song" at Doc's funeral in the upstairs sanctuary of the Epping Community Church. Toward the end of the bizarre solemn event, the widow announced that Doc had asked her: "Sheila Honey, you sing so pretty! Please write me a funeral song and sing it so I can hear what it will sound like when I die."

Can you IMAGINE?! Doctor LaBarre never asked her to write a song like that! He was abused. Not crazy.

As she sang in the church, my eyes twisted into a knot inside their closed lids. 'Caterwaul'. That's the word for it:

"to make a shrill howling or wailing noise." She keened on about how "Eye-eye-eye will mi-i-i-iss yoooo-oooo-oooo so much when you're go-o-o-one!!" She went o-o-o-on and o-o-o-on and o-o-o-on, wavering up and down through the whole wretched delivery.

This performance followed an equally edgy parade of chiropractors who came from all over the country to express appreciation for money Doc had donated to their programs. This was after Sheila had explained to the hundred assembled mourners that the deceased had broken his own neck while giving himself an adjustment. UNbelievable! No wonder the chiropractors looked so worried! They KNEW it was hideous and what must they have thought of Epping?!

Following our friend's death by self adjustment, Sheila had a steady rotation of "menfolk in and out," as Bruce Allen's mother described it to a Boston television crew.

Big news on the farm.

Sheila had apparently murdered a 24-year-old Massachusetts man and burned his body on her front lawn. Then, she chucked his remaining bones and melted mock Rolex into three garbage bags and tossed them into a compactor at the town dump, a testament to our efficient trash processing. There was some glitch about the dump being closed on Monday. Hence, the untidy burn pile. A second set of human bones was also found on the farm. Plus nobody knows where another guy went, but they found his toes in the septic tank.

You can appreciate that we feel real good about how that former-Harvey property has gone down the tubes. During Sheila's murder trial, a stream of realtors and easily-recognized developers drove through our and Allens' farms to the dead-end LaBarre property.

It was the oddest thing to see both of my parents on Boston Channel 7 commenting, along with Fran Allen, on our widow woman's behavior.

"Oh yes," said my mother. "We've seen a lot of them come and go since Doctor LaBarre died. All young fellas in their 20s, I'd say. Sheila seems to collect 'em." Her "collect" sounded more like "clucked," her eyes appearing slightly mischievous through the eye glasses she has worn all her life.

"We've even had some come down here, all beat up," she continued, apparently at ease with the camera man. "I remember this one fella, this was about five years ago, he came in here, cold and dirty and smelly. We gave him some tea and a sandwich. After that? I don't know. We never saw him again."

And Dad, struggling not to squint under the bright lights, said: "You haven't lived until you've met Sheila!" The S.P.C.A. rescued the horses, which Cindy eventually got from there.

Persistent reporters even got the reticent farmer Bruce Allen on camera! His gaunt weathered face appeared even grayer than in real life, as if recorded with black-and-white film, and his voice cracked as he described himself as "uneasy" about "goin's on" up the road that passes through his barnyard. As he turned abruptly to walk away, the guy with the microphone caught a good one. Bruce muttered, "If you knew Sheila you'd be uneasy, too!"

I'm still trying to understand why someone had to die before our dominatrix neighbor could be dealt with. It seems that everyone had heard about, if not actually experienced, her "Medical Room," a claustrophobic sterile-white chamber furnished with a gynecological table, a dentist's chair and a toilet. They said her laboratory cabinets were well stocked with "toys." And, during a week of police searching at farm, ours and Allens', it was revealed that most of the local men – including Dad – had seen more than they expected to see of Sheila when she opened her door dressed as a slutty nurse in tall white patent-leather boots and a see-thru pharmacist-style shirt with buttons along one shoulder.

This had gone on for years. The bludgeoned boys slouching through our farms with their torn shirts and black eyes, the neighborly exposures. And, then Mom so casual in her telling about her numerous interactions with forlorn farmhands fleeing the hill, as if we take these sorts of things for granted, which, I guess we do. Or did before the shit hit the fan.

It's nothing to be proud of – our ignorance of the neighbor's abusive relationships with young men. By 'ignorance', I mean that lots of people in town, including the police, looked the other way. Like when a thief stole a toilet right out of the police station, they had nothing to go on.

My family seemed to accept the situation with a live-and-let-live attitude. Or could it be that that Doc, Sheila, and her consenting companions had become so familiar to us, a part of our scene, that we were fully aware that the conditions in which murder would be committed did not apply to us? Whatever the game was, we weren't playing. Sheila LaBarre could continue to visit and not hurt us.

Would it be different in a city – if we all had separate apartments and a security guard at the door? Would that make us more cautious and alert? Or less? I don't know.

When they caught Sheila in Massachusetts, she had altered her appearance. First described in media alerts as a "blonde with brown roots," she had become a bouffant redhead. Everyone said she looked "so good" for her arraignment.

But we knew she wasn't.

If my family had retained ownership of that farm we would've at least fed the horses and kept the whips and chains in the stable.

Slippery Sheilas. There was always someone wanting a piece of the kingdom.

Part Four – To Protect The Innocent

The way we were ...

35. TO PROTECT THE INNOCENT

Note: As is often the case in life, this chapter is pretty much bullshit based on actual events. It is the first thing I wrote when beginning my story. I meant to write a novel because I thought it would be easier and I had never tried writing fiction before. So, I changed all the names and put Dad in jail. But he wouldn't stay there – as you will see.

On the morning of March 15, 1983, my father's friend Bob Button was hauled from our farmhouse, off to jail. A Rockingham County sheriff's car pulled into the gravel driveway, followed by an Epping cruiser with Chief Lodge at the wheel. Reporters and photographers from three papers had already arrived. A TV crew was set up on the doorstep.

We had been hiding Bob for two weeks – though it wasn't much of a secret, with Libertarians coming and going, making a spectacle of themselves. But now the jig was up. The Epping police were knocking on the door with a warrant for Bob's arrest.

Hold up a second there!

OK. This is the chapter that is the most "almost true." It wasn't QUITE like that. I just haven't found a better way to sketch Dad's character as a notable heckler of local government's attempts to adopt land use regulations. Events described here happened, but not all at the same time or around the same place. Aside from sentiments ringing true, this chapter is fiction.

People who know my father can tell their own stories about how he is. If you are one of these relatives, friends or acquaintances and can stitch together a better profile, I say: 'Go for it!'

Dad's friend Robert Bonser DID go to jail for refusing to get building permits for mobile homes in his 'Cedar Waters' nudist park in Nottingham. Epping police cruisers and TV and newspaper reporters HAVE come to our farm many times over the years. And my father IS NOT slow to defend his and everyone else's property rights, even if everyone else doesn't care that much. And Daniel Webster

Harvey HAS TRIED, unsuccessfully, to defend himself and the United States Constitution in courts of law. He has helped push at least one attorney toward a nervous breakdown. And my Uncle Eddie HAS tried, repeatedly, to talk sense into the Old Boy all their adult lives.

Obviously, I think a lot of Dad and my Uncle Eddie – even if I'm apt to lie about them from time to time.. Ever-present in my life, Uncle Eddie strives to do the right thing while Dad considers it his duty to lean the other way. Ed is the most documented person I know. He has kept a detailed journal most of his life, recording events of the day and things like what he ate for lunch and how much fuel his bulldozer consumed. So, he'll know, without doubt, where he was on the Ides of March 1983 – probably nowhere near our dining room table on Red Oak Hill.

Julius Caesar's soothsayer was no jester when he said: "Beware the Ides of March!"

Duly warned, I continue this almost true chapter with fictitious names, dates and hair color changed to protect the innocent, acknowledging that it was Victor Porter of West Epping who said in 1978: "If I had wanted one of those, I'd've bought it before now ..."

At our house, Dad had been briefing the press since 8 a.m. Bob Button showed few signs of life as he slouched at the table. He said he was eager to go to jail, but wasn't what you'd call 'enthusiastic'.

Reporters and a few townsfolk stood in our kitchen, coffee cups in hand, grease-stained boxes of donuts on the table. Cousin Walter, (now a barber in Ballarat, Australia) worked in a bakery in Raymond before showing up for school. Though we didn't know it at the time, my 16-year-old cousin had been paying attention and was destined to leave Epping just as soon as he graduated. He anticipated our need for donuts on the morning of Bob's arrest.

Some of the crowd spilled into the adjacent hallway and dining room. Bob's bags were packed by the front door. Bob's wife, Bertha, had given up on him two weeks earlier when he was held in contempt of civil court. Apparently, she found him contemptible, too. She went to live with their son in Lee – a move from one trailer park to another.

So, it was actually my mother, Louise – not Bertha – who had packed Bob's bags. She ironed flannel plaid shirts for both Dad and Bob and laid out clean underwear and jeans, even made my brother Matthew polish Bob's shoes so he'd look respectable enough for prison. (Note: This REALLY didn't happen. Matthew doesn't polish anybody's boots.)

"These MEN!" she was heard muttering, off and on, during the past few days. Sometimes, she simply uttered: "LIBERTARIANS!" as a one-word damnation of all the trouble Dad had gotten us into during the year. Mom knew – and I witnessed it – Dad had been "carousing with" political activists who quoted The U.S. Constitution and Bill of Rights with reverence similar to Fundamentalists citing The Bible.

Libertarians were stirring up resentment and clever strategies in Epping and neighboring towns. They met secretly at the 'Route 125 Restaurant' in Lee after it closed for the night. The eatery owner was a sympathizer suffering from building permit frustration. Sometimes the rebels met at cohorts' homes, always with the understanding that parking would be discrete, not all in the driveway.

Mom's cross-stitched sampler was framed over the gas range. Under glass to protect it from steamed tomatoes, it declared: "No Matter Where I Serve My Guests, They Seem To Like My Kitchen Best." But Louise wasn't very hospitable the morning of the arrest. She just sat there by the sink behind Dad and Bob and didn't say much. She didn't have to. Everyone else was talking.

Especially my father. He was Bob's handler, his spokesman. It was Dad who convened the pre-arrest press conference. He had called the reporters.

"He said he'd be here," my father was saying. "But I told him we can handle this. When it comes to property rights, we don't need a lawyer to speak for us. Do we Bob?" Bob shrugged and smiled unconvincingly as a camera flash went off. Dad referred to someone named "Melanson," a free attorney dispatched from Fort Wayne by the Libertarians.

Reporters scribbled in their notebooks, two of them seated at the table, one standing. Dad was obviously pleased to have his words carefully recorded. He waited until the scribes caught up. History in the making. That's what it was! Going public at last with the truth about the unConstitutionality of building permits.

This was the day Dan Harvey would have his say and prove his support for the Little Guy: Bob. It had been a long time since Dad made front page – not since he ran out of gas and landed in Boston Harbor. This would be good! Dad's squinty eyes twinkled with mischief.

What had he told the judge two weeks ago? "Your Honor, if my client, Mr. Button, wanted one of those, he'd've bought it before now ..."

Bespectacled Judge Turcotte had leapt to his feet, robe ascending and bailiff bleating: "All rise!" Why was he wasting his time with these clowns? Hadn't he just spent an hour in chambers with attorneys for prosecution and defense – that Libertarian Melanson – and Harvey for the defense? Hadn't they agreed to settle without penalty if the defendant would simply go through the consent process? Hadn't they JUST agreed?

"Mr. Harvey! Mr. Button!!" admonished the judge. "You are both an insult to these proceedings! I'm giving you two weeks, Mr. Button, to comply with this court's recommendations. You need to buy a building permit. If you don't, you will spend 30 days inside the county's correctional facility. Do you understand?"

Bob understood. So did Dad, who would have gladly traded places with his friend. But Dad couldn't get this attention on his own – though he had tried. He had purposely neglected to obtain building permits for several farm buildings – major ones, like the milking parlor and pole barn. But neither the selectmen nor planning board would take the bait. They ignored Dad, refused to fight, said these were farm buildings and didn't need permits, so settle down.

Bob's wife Bertha was next to stand in the courtroom, and it wasn't beside her beloved. She probably understood Judge Turcotte's order better than anyone else in the room. She knew EXACTLY what Bob would do before Bob knew himself. And it was too much for her. Bottle-blonde Bertha headed for the door where she shoved past the Exeter News-Letter reporter and a photographer trying to take her picture. She marched to the parking lot, yanked open the door on Bob's Impala, slammed it into gear and drove home to grab her robe, toothbrush and slippers.

Both my father and Bob Button are substantial men, the handsome big-boned sort of achy-muscled farmers that most towns suffer. Each six-feet tall, they could be mistaken for brothers. Between them they'd have a full head of hair and a complete set of teeth. Dad is known for startling waitresses by pulling pliers from his pocket to extract his upper plate and asking it: "Are we ready to eat?"

But he wasn't fooling around on the morning of Bob's arrest. He played a different role: that of independent thinker and champion of the 'under class' and the 'common man', defending his 'Little Guy' friend Bob. Dad leaned forward to engage the reporters, his arms resting on our Formica table. He told his audience about the Boston Tea Party.

"If I was around then, I'd be one of those fellas," he declared.

All week, we Harveys had anticipated this scene, another coming to terms with Dad's politics. It was his recreation. He'd act up and act up some more, spend expensive time with attorneys, thumb his nose at selectmen in three towns where we owned land. Tension built. Authorities got tired of being harassed or ignored. And, well, it would come to a head – a court date, a registered letter, a summons, a lien on the property – or, in this case, an arrest.

The previous day had passed with more and more people 'just stopping by' to see Dad and Bob. Several friends had tried to talk sense into the 'Old Boy'. That's what most people call my father, though he prefers the title of 'King'. A bunch of us family had been in and out of the house all day. We took turns sitting on the porch with our two hired men and a slew of Libertarians.

Dad didn't stray from his chair at the head of our long oval dining room table. It's where he presides over Sunday dinner. Mom sawed pine boards for extra leaves to make the table long enough for all of us. Three tall windows face southeast. Plenty of sunlight immediately cooled by blues in the room – the bright blue wallpaper and soft blue tablecloth, the cobalt lead-light hanging lampshade that my brother Matthew made from a kit. An oil painting graces one wall – golden fall foliage landscape, stonewall-lined road with horse and buggy. It's a pleasant dining room full of good memories, a fine place for Dad to hold forth.

That day before the arrest, Uncle Eddie made an appearance. Of all our relatives, he's the one most concerned about Dad's refusal to play nice. We watched as he got out of his pickup donning his familiar leather Australian bushman's hat. The burly man's confident stride across our driveway became more erratic when he nearly stumbled over a heap of crushed stone by the door. Hard not to notice a dump-truck load of crushed granite, but my uncle walked right into it. The sun must've been in his eyes. Mom buys crushed stone as a gauntlet to keep manure and slugs out of the house. It cleans the bottom of boots and hurts bare feet. Crushed stone notwithstanding, Uncle Ed assured us that everything would be all right. Said he would talk with 'Daniel' and entered the house like Christ ready to raise Lazarus.

I followed him, hoping to witness a miracle. Hat in hand, blue suspenders holding up his pants, my uncle gave Mom a sympathetic nod as he passed her in the kitchen, headed for the dining room.

"No-n-no. Don't get up," he said to Dad who made a token half-effort to rise from the table.

"Ahhh, Ed! Are you a friend or an enema?" my father responded. "Pull up a chair. Take a load off."

My uncle spoke solemnly. He pretty much always speaks solemnly. I think it's because he climbs the Community Church steeple each week and winds the clock. (My cousin Dana still does.) Uncle Eddie appreciates reverence and respects time. He takes his work seriously and writes it down in his diary. Once, I asked him once how often he checks back. He said the journal came in handy when he excavated a golf course and one of the other subcontractors tried to sue the developer. After he had been sacked, the plaintiff wanted back pay. My uncle referred to his notes and testified that the slacker contractor had taken days off and his coffee breaks were too long.

Though Dad writes a lot, too (on yellow-lined legal pads stacked in the bathroom), he was not pleased with his brother-in-law's paperwork on the morning of Bob Button's arrest. You see, Uncle Eddie had recently gone over to the Dark Side. He had joined the Epping Planning Board and helped draft 75 pages of zoning regulations. He had also been 'cavorting' with the selectmen.

"You know, Dan. We all understand what you're trying to prove here," my uncle began. "But you shouldn't encourage Bob to break the law." Ed said he and the selectmen (and the unmentionable planning board, no doubt) admired Dad for his commitment, for his spirit of Yankee independence, his defense of Bob.

"But ..." he moaned toward his plea for my father to be "reasonable" and consider his family, "our" family and Bob's family. "You're shooting us all in the feet, Dan. You know there's no escaping the planning board and building permits. It's NOT unConstitutional! It's PROGRESS. And, you're making the TOWN look bad with all this foolishness!"

Ed said he was sent by Epping's three selectmen, the "Town Fathers" or "Three Wise Guys," as my father calls them. "They asked me because I'm your brother-in-law and maybe we can work something out. So, well, I'm asking you, Dan. Please! Please don't force their hand. Just tell Bob to comply and buy the building permit. He'll do whatever you say. Don't let John Lodge take him away like a common criminal. It's no good! Please, don't be so damned stubborn!"

It was, of course, too late for Ed's entreaty. My father was already damned stubborn and Bob Button was committed to becoming a Constitutional martyr. Louise had done a great job packing his suitcase. At last, his life looked orderly. His future felt accessorized, clean and well folded, ready to move on.

"Thank you for your concern, Ed," my father said. "But this is what we believe in. American freedom as vested by the Bill of Rights. It's a sacrifice Bob wants to make. Heck! I wish I could take his place. Harveys have been here since 1755! Can't you understand what that means? Why should THEY have control over ME? Or any of us?"

"So be it, then," Uncle Eddie said, making his dejected exit. "I'll tell 'em downtown. And, I hope you grow up ... sometime."

Next day, they took Bob to the Rockingham County Complex. It was on TV and in all the papers. Soon over, though. Quieted down by the weekend. Nobody seemed to notice when Bob got out a month later, hoping Bertha would take him back. While he was in jail, the Town of Nottingham removed five mobile homes from his 'Sticks and Stones Nudist Park', dragging them a half-mile down the road with fish bowls sloshing, skirting and porches flapping, to where they sat in a sand pit for nearly a year.

Cousin Margaret Olive Woodward Redhouse and my youngest sibling Erica Alice Harvey Martel.

My Brother-in-Law Tom Orzechowski, married to Susie.

Melanie Fike and Otis Clapp were married June 15, 2007 at Camp Menatoma, Readfield,

The way we are ...

36. HAY WARS

The Sisters' Refrain ... (please)

There once were six sisters from Epping
Whose spirits all needed some pepping,
While two followed tradition,
The rest threatened perdition,
Sisterly bonds overstepping.

* * *

Farmerettes insist on their way.
It's us who still pay and pay.
Agriculturally speaking
We think they need tweaking.
They can't make believe it's okay.

Last summer, I sort of got suckered into doing Phyllis's haying, just a little of it. Or at least I thought I was haying for Phyllis. That was the sucker part.

Home from New Zealand and trying to fit in again, I was flattered when Erica and Cindy invited me to join their haying crew. Cindy said I could rake. "Go get on the tractor in the Orchard Field and rake. I'll be along in a couple hours with the baler. It'll be fun, just like the old days!"

She said Erica and Mom had bought a new round baler and they ("we") were taking over haying because Phyllis and Beth couldn't do it right. She said Phyllis's hay conditioner and baler were pieces of junk and, anyway, she and Beth might hurt themselves.

"So, we're just going to do it FOR them," Cindy declared.

"Does Phyllis know about this?" I asked.

"What's it matter? The hay's down. Go rake it!" she said and drove off to watch her daughter, my niece Abbie, play volleyball.

I wasn't sure how these haying arrangements would work. Seemed like the three sisters (Susie, Erica and Cindy) were united again against "The Farmerettes" (Phyllis and Beth) in an agricultural takeover.

Phyllis said 'The Sisters' had no right to be so pushy. She said they were making our parents go too far. "It's unauthorized, what they're doing!" Phyllis declared.

But how could it be unauthorized if Mom was in cahoots with a new baler to bundle new-mown hay for storage? The nod of Mom's endorsement tipped my scale. It was a hot afternoon and I didn't have anything else to do, anyway. Besides, it felt good to be asked for help. Cindy had even flattered me by assuming I remembered how to drive a tractor.

She was right. It was, as she promised, 'just like the old days' as I figured out the gears and roll-started a John Deere Model B from where it had been perched on a familiar hill. The side-delivery rake rattled and squeaked, combing the hay into windrows, which I doubled (two rows raked together) to save time baling. I hadn't seen the new baler, but surely it could gather a double row.

The two hours passed quickly because it only took me an hour to finish going round and round the Orchard Field. I totally enjoyed the experience and wished I could do more. It was fun being whipped to and fro by the tractor as I gripped the wheel and kept an eye on the contraption behind me. The sun burned my arms, but so what. There's joy in riding an iron beast, passing through the exhaust and dust, chugging and clattering across the field. I felt healthy and strong, grateful to be doing something on the farm.

Cindy returned, according to plan. She drove another (newer model) John Deere pulling the baler that gathered and rolled the dry hay into toilet-paper-shaped bundles weighing about 300 pounds each. It didn't take her long to scatter 50 of these bales throughout the hayfield.

Later, I joined Cindy and Erica plus Phyllis to roll the bales into a Night Pasture box trailer for storage.

Although The Sisters disparage Phyllis's ability to run farm equipment, she is the most capable among us – probably not as good as Matthew or Albert – but best among the girls. For years, she actually earned a paycheck away from the farm as a contracted equipment operator, running a loader in a gravel pit filling dump trucks. Plus, she spent one summer working nights to prepare Hampton Beach for tourist sunbathing. She managed a 'Salamander' to sift sand.

Another of her jobs involved servicing home septic systems, sucking effluent into a tanker.

So, for me, there was nothing odd about Phyllis using a loader to shift the heavy bales. The process should have been sweet with four of six sisters pulling together in a rare display of teamwork. Instead, the tone was edgy.

Erica, Cindy and I were in the trailer. Phyllis was outside operating the Bobcat.

"Look at her!" Erica said. "She won't get off that machine for ANYTHING."

"Yuh. You'd think it was stuck to her rear," Cindy said. "And she thinks she can do this by herself!" They said they were tired of watching Phyllis and Beth "screwing things up" and breaking the equipment.

I didn't get it then. Or even a week later, when Erica complained to me that Phyllis was "feeding that good hay we put into the trailer to CALVES!" She thought Phyllis should save it until winter for the milkers. She didn't think the calves needed such good feed and suggested that I intervene and tell Phyllis to leave the good hay alone. I said it didn't worry me and it wasn't any of my business. Weren't 'we' giving it to Phyllis for the cows?

Apparently not. Six months later, when I returned from New Zealand for another summer in New Hampshire, Phyllis said 'The Girls' took most of last year's hay to Cindy's barn and sold it from there to horse owners who paid extra for quality hay – more than Phyllis could afford. She said she was lucky to be able to buy 200 bales from Rockingham County Farm in nearby Brentwood.

I told her I was surprised it had worked out that way and wondered what factors prompted The Sisters' decision to demand payment for the crop that Phyllis and Beth had expected to harvest themselves.

Could it have been The Farmerettes' alleged incompetence with field crops? (As far as I know, nobody in our family, except Matthew, ever took a course about growing hay and corn. And, I must say that after seeing how seriously New Zealanders take intense land management, I wonder why we Harveys work so hard at doing such a half-assed job with the crops.)

Withholding hay was no surprise to Phyllis.

"Mommy and Daddy have been watching them roll over me for the past two years," she said. "People say how wonderful we are, how Susie and Erica take care of everyone in town. But they're just mean to me and the cows. You can keep my share of their sisterly love!"

When I asked Mom if there was a 'hay war', she barked: "Why SHOULDN'T Phyllis pay for the hay?!" The hot-button response startled me. Her words were pre-loaded, a quick cut to the chase. I clarified that I had assumed Erica and Cindy were doing Phyllis's haying for HER last year."

"Well. You were wrong," my mother said.

Later, I asked our son Amos for insight. "Grammy is very touchy about the hay," he said. "When she asks: 'WHY shouldn't Phyllis pay for the hay?' she's really asking: WHY? She wants to know WHY Phyllis can't afford to buy the hay. You know. Like WHY isn't she making any money by milking cows?"

Amos also reminded me not to forget what happened two years previous when his cousin Danny, with help of sisters Analesa and Sharon (Matthew and Judie's kids), did their usual good deed of filling their grandparents' barn with baled hay.

The intent was for my parents to sell the hay throughout the winter and use the money for personal expenses. It had worked that way during the previous several (not sure how many, but more than a couple of years) and had been a sweet routine of the kids working with their grandfather. Dad ran an old baler, as he always had, and Danny's crew did the rest.

BUT there came the year when Dad stood in the way of Danny's mid-winter negotiated hay sales. He told Danny not to bother selling the hay (to benefit my parents) because "it was spoken for." He had GIVEN it to Phyllis for the cows.

So, yes. It's a good question: Why SHOULDN'T Phyllis pay for the hay?

With all the resources available to her, WHY can't she make her own hay? Or at least have enough money to buy it? Sort of gets to the crux of the problem, I think. Made for a lot of hard feelings within the family. The grandchildren were not keen on supporting Aunts Phyllis and Beth, instead of Grammy and Grampy. For their part, my other sisters remained grim and humorless while watching the routines and relationships unravel.

Part Four – To Protect The Innocent

We used to get along with each other better. I keep wondering if it's 'Boiled Frog Syndrome'. Like when you put a frog in a big kettle of cold water, then slowly heat to boiling. They say the frog just stays there and won't jump out unless someone pokes it because the slowly warming water numbs the victim into thinking nothing bad is happening. Waltzing back from New Zealand, I find the Harvey water coming on the boil and nobody notices because everyone is too stewed to jump out.

So I tend to keep my head down and keep working, out of town if possible. I accept that my prolonged absences void any family privilege I may have enjoyed in the past. The hay war is really none of my business.

Forget the new baler. Forget the tension. Forget the struggle.

Did someone claim that milking cows would cover the cost of overhead, feed and taxes?

On Red Oak Hill, at least, it never has. Through borrowing, our parents kept the farm afloat until it sank in 1987. They were on the hook to three banks for half a million dollars ... WAY more money than Phyllis has invested. I think all she owes is $8,500 for the manure pit. It would have been a lot worse for her if Dad didn't cover the financial slack.

WHY isn't Phyllis making money? Why doesn't ANYONE who milks cows in New Hampshire turn a profit?

Submit answers to P.O. Box 578, Epping 03042.

Beth christening an experimental windmill on top of hill, 1978. Witnessed by state dignitaries and Colorado engineers.

Erica and Matthew serviced the mower-conditioner.

Red Oak Hill was selected to site an experimental tower and wind machine as part of a New Hampshire 'Clean Energy' initiative. Intended to power Matthew's nearby house, it started off drawing electricity from it. Rockwell International flew an engineer from Boulder to rectify the situation by flipping switch.

Grammy Eunice Peterson in car (but you can't see her). Dad, Mom, Susie, Otis and Amos.

The way we were …

37. ONASSIS POLITICS

In 1973, my father considered overseas options for making money. During a blitz of bizarre press coverage, Dad offered a deal to prominent Greek shipping magnate Aristotle Onassis! He tendered ocean-front land he had inherited from the Adams Family to Onassis for an oil refinery.

Conservationists throughout Rockingham County and the Rye Board of Selectmen spoke out against the proposal. Dad sat by the phone for two days, but his offer didn't amount to more than fleeting headlines in 'The Exeter News-Letter' and 'Portsmouth Herald'.

At that time, there was a buzz around the man who paid Jacqueline Kennedy $3 million to marry him. Onassis taunted nearby Cape Cod (Massachusetts) residents with plans for an offshore super-tanker docking facility. Things quieted down when the tycoon's agents and engineers abruptly left Boston. They had not been serious about buying our Rye land. It was just a ploy to strengthen their stance in Cape Cod.

During that decade of financial instability and encroaching humanity Dad spun like a cornered porcupine. He preached that "civilization as we once knew it" was coming to an end. For him, it probably was. He visibly strained under the weight of unstoppable land use regulations. His inheritance and recent investments in real estate were no longer solid. The government chewed holes in his dominion. His authority was slipping away.

My father hates to be told what he can and cannot do on his property. Three demons haunted him after the Epping planning board was created in 1968:

1) The Master Plan,
2) Zoning, and
3) The Loathsome Planning Board itself.

He vigorously opposed the inevitable ballot-box adoption of all three and was a real stinker about technicalities. He insisted on re-counts after each vote. And even though Epping's zoning and building 'Codes' book is now in its 42nd printing, Dad still doesn't believe in building permits. (I have obtained three now, much to Dad's dismay.)

Entering the 1970s, my father had a lawsuit going with the neighboring town of Nottingham over the size of house lots. Not that he intended to sell anything, he tried to preserve his 'right' to build on one-acre houselots by overlaying 200 acres of forest with a checkerboard subdivision grid. He anticipated that the Town of Nottingham would adopt a code prescribing a two-acre minimum lot size.

So, each lot defined in Dad's virtual subdivision was just one acre in size. He proceeded to deed every other lot to every other kid (just we three oldest: Susie, Matthew and me). That way no two sections could be amalgamated because none of the owners owned adjoining property.

I'm not sure whether or not Nottingham ever imposed the two-acre minimum requirement, but there was an uproar over Dad's pre-emptive strike. And we ended up in court. Despite Dad's evidence proving that the planners had failed to advertise public hearings on time, the Superior Court judge ordered my father to void his clever subdivision.

Dad was correct in identifying 'irregularities' in the Nottingham Planning Board's procedures. Still, the judge said a volunteer board can't be expected to do more than 'substantially comply' with the law and Dad should just suck it up. Losing the case was discouraging. But implications for future contests were even worse. Dad's case has since been cited as precedent against citizen pleas against other volunteer boards where letters of the law weren't followed. Much to my father's chagrin, he may have codified "substantial compliance."

Chastised by Nottingham, Dad joined other soreheads to publish planning-board-bashing newsletters. A group of Epping's 'concerned citizens' stuffed mailboxes with political flyers and cartoons. They called themselves the 'Epping Betterment Association' and held weekly strategy meetings. Dad said they were "plotting the overthrow." Their mission: Block land-use controls.

Several members of this group were disgruntled land developers who assumed Dad shared their ethics. They urged him to develop the farm while he still could. They were not farmers and did not understand Dad's motivation to keep land open. He was political, not entrepreneurial. His association with the E.B.C. did not endure.

Still, Dad bristled at those in authority and was not at all happy about the way things were going.

New Hampshire's population had been growing steadily, indeed at an alarming rate within southern counties along the Massachusetts border. Epping lies at the crossroads of two major highways (Routes 101 and 125) within easy commuting distance of industrial areas over the line into Massachusetts. The neighboring "Bay State" was reputed to offer higher-paying jobs than New Hampshire, while we boasted the lowest taxes in the entire United States. Judging from the volume of traffic headed south in the morning and how quiet Epping was midday, the tag 'bedroom community' fit.

Of course, land values rose as people began paying higher and higher prices for house lots.

New Hampshire still has no income tax and no sales tax, no 'broad-base' tax like other states. There is no bulging bag of money held by the state to fund public services – like schools, fire, police and trash pick up. Therefore, each town is pretty much on its own to collect revenue to pay the bills.

With more and more families moving into town from "Taxachusetts" during the '60s, New Hampshire's small towns often spent as much as 90 percent of their total municipal budget on public schools. This made it difficult for Epping to retain state accreditation. Most voters didn't even have kids in school. And, if they did, many sent them out of town for a better education.

Each year, starting with seventh grade, many of my classmates were 'tuitioned' away to Exeter, Kingston and Durham. They expected extras like 'physical education' and 'home economics' and fancy languages like Spanish, German and Latin. David Bascom was tuitioned to Dover because he wanted 'shop'. A couple other guys traveled to Manchester to be Catholics. But they came back.

Property taxes – assessments on buildings and land – were the primary source of any Granite State town's revenue. Larger towns could tax businesses and industry but, back then, Epping had just three mom-n-pop grocery stores, a skeleton-crew brick yard, and a shuttered shoe shop. Real estate was reasonably priced in towns like ours and attracted 'first-time home owners'.

Thus set the scene for New Hampshire farmers to squirm and fall for the temptation to sell land for house lots. The financial incentive were twofold: They could get heaps more money for developing than for field crops, pasture or forest land; and 2) The towns were assessing them for house lots anyway to pay for the schools, so why not sell quick and get it over with? Mark that one SOLD!

Tax breaks through current-use assessment were coming into play at about that time. Still, there remained questions about the long-term effect of enrolling in the program. State and local governments were still trying to figure out how to balance budgets.

Many landowners leapt when legislation opened the window for speculation. They did not bother with current use assessment and simply sold what they had for the highest price they could get..

We Harveys weren't that inclined to sell out – not for housing developments and certainly not to the government for a deal that Dad hated. He said tax assessors should appraise property for its current value without making people sign a tax-break contract. He said we should be able to sell land to whoever, whenever, for whatever money we could get. Just the way it had always been, minus speculative tax bills.

No. We didn't intend to sell anything. We were happy to continue farming. But what if we wanted (or needed) to sell a few acres? If we were encumbered by a current-use agreement, then: 'Oh Dear!' Begging to get out of it would be too big an affront to Dad's stature as a responsible landowner. He was slow to cooperate as high property taxes continued to be our biggest worry.

Albert with sons Nathan and Ned

The way we are ...

38. SUICIDE COUNSELING

The Sisters' Refrain ... (please)

We wish we could help them, we said.
We know there's a rock in their head.
Is it beer or drugs?
We'd like to give hugs
But we'll send them to therapy instead.

* * *

It's enough to get us unglued
Tractors run by incompetent dudes.
The barn needs new wiring.
It's got us perspiring.
It's a wonder we haven't been sued.

Here's a good one!

Remember how Phyllis failed to write her 'Last Will and Testimonial'? Well, there was another appointment that went even worse. Phyllis said Susie came to her "pretending to be concerned about Beth," asking Phyllis to help get our 45-year-old sister some counseling because it wasn't 'normal for her to be so mad about everything'.

"Susie told me there was this free service and I should tell Beth 'C'mon! It'll be fun. An afternoon on the town. And it's FREE!'" Phyllis told me.

So, Beth agreed and they showed up at an office complex somewhere in Portsmouth.

"We told the receptionist that we had an appointment. She gave us questionnaires asking about our suicidal tendencies," Phyllis said. "So I asked her what that was about and she went off to get a doctor."

The doctor and nurse came out and took my sisters into a room.

"You ARE here because of suicide, aren't you?" the doctor wanted to know. Phyllis told them 'No'. They were there because it was FREE.

"She came back and began: 'Well, your sister says ...' and I knew Beth would hate ME for tricking her like that, all because of Susie."

Phyllis said the counselor kept them for an hour, asking why they weren't dressed better and did they have drugs in their handbags? "I don't even have a handbag!" my sister laughed. "He asked: 'What IS that?' Looking at the zip-lock bag I carry my wallet in. He didn't like our work boots or our hair.

"We weren't very cooperative because Susie had tricked us into going. We hadn't even tried to kill ourselves. Or anyone else! And, you wanna know the WORST part? Beth keeps getting a bill from them for $120!"

Once again, I asked my sister WHY she even went along with such a scheme. "Why did you even go?" I asked. "You know The Sisters are gunning for you. They say you're hiding behind the cows at the farm, not going out and not paying your way. It was just a set up."

Phyllis shrugged and said: "They're probably right. It's awful hard to make ends meet. I just try to mind my own business and get along. But do they have a right to evict me from a farm that I have a share in? Was a vote ever taken? I don't think so!"

"Well," I began, not wanting to go much further. "You know Susie calls the shots around here. You all say she does whatever she wants and I'd better get used to it."

Phyllis wasn't eager to go further either. "Yuh. And Erica and Cindy are right behind her. But do they have to go telling people that Beth and I are 'the village idiots'? Why should we have to put up with that?"

I told her I was sorry.

Truth is, I wasn't surprised that it had come to name calling and tricky counseling. Erica had already told me that she and Susie and Cindy were genuinely concerned about Phyllis and Beth's behavior and well-being. In addition to not paying Epping taxes, she said the two 'farmerettes' were managing the dairy and crop work recklessly, letting "anyone who came along" run the machinery. She said there might be an accident and we could lose everything.

"Besides … They're not making any money," was the refrain.

There were two sides to every sad story. I had already heard complaints and explanations about why electric fences were dragged away one day and put back up the next, and why rows of corn planted on top of each other were not necessarily proof of drunkenness. A tractor had rolled over, but the man who had volunteered to drive it was not hurt.

Phyllis and Beth said they couldn't do all of the work by themselves. Hadn't the farm always taken advantage of willing workers, anyone "bright-eyed and bushy-tailed" eager to learn about farming? Why couldn't our family be more helpful? More supportive? Why did Albert always expect to be paid when he repaired the equipment? Why did Daddy keep telling them how much better the McPhees were? Why did Mommy scowl at their friends?

It was getting to be too much and – being good Harveys – everyone had an opinion about who was at fault. And it looked like Phyllis and Beth were outnumbered.

I recall going to aerobic classes with Phyllis twenty or so years ago. She was 'Miss Personality' in her pastel-hued leotard and shorts, punching and prancing to the pop music, humorously heckling the instructor. The class loved her.

I called her "Flower," short for "Phyllis The Fair Flower." And, now ... well ... she looks more like ... a farmer. Worn out. I don't care what they say. Hard work outdoors in four seasons does the body no favors! Phyllis's current wardrobe is far from leotard and shorts. She wears unflattering manure-and-grease-stained coveralls and has let her hair go gray and straggly under a ball cap.

I'd just like things to be the way they were when we were all young and good-looking with our lives ahead of us.

Is that too much to ask?

Louise, Dan and David in our '51 Plymouth, another party on hill. John & Heather Fernald's wedding.

Historically speaking ...

39. N.H. DAIRY INDUSTRY – DON'T HAVE A COW

According to sources cited by the Cooperative Extension Service for 1996-1997 ... (Stale data. I've been working on this project TOO long!) ... it's worth $18 billion dollars a year to New Hampshire's economy to preserve its then-identified 5,265,000 acres of open space. ($1.8 billion of that collected as state and local taxes.) Plus, they say over 200,000 Granite State jobs depend on scenery-related activities such as farming, forestry, tourism, recreation and maintenance of vacation homes – more than all that now, no doubt, as we are into the next decade. Surely, these numbers have grown.

However you figure it, these are no small potatoes and pressure to develop land is increasing.

City and town officials often consider undeveloped real estate to be lost tax revenue, especially now that nearly all of it gets a tax break through current use assessment.

Municipal Budget Committees suggest that more tax income would be generated if open space were converted to higher-priced developed use. BUT recent studies cast a different light on the situation. Researchers calculated potential high assessments for various land uses (residential, commercial and industrial) and subtracted the cost of services these uses would draw down from the annual town budgets (for schools, fire and safety, water and sewer, highway maintenance, etc.).

In every New Hampshire town studied (as of 2008), open space pays more in taxes than it requires in cost of public services. AND, in every town except the lakeside community of Alton, the services required for residential land use cost more than taxes paid by people using that property. So, duuhhh! Open and agricultural land doesn't generate big tax money for a town when it is assessed as it should be – for its current use. BUT neither does it cost as much as development would for services.

Dad tried to tell them that 45 years ago!

Here's another revelation: Most people would rather admire and hike through open fields or forested foothills than walk through a housing development or mobile home park. That's why open space is desirable and of public interest.

Nevertheless, New Hampshire's working landscape continues to vanish as large tracts are developed to meet the demands of the fastest growing state in the entire Northeast United States. Older dairy farmers, in particular, are increasingly discouraged by always-low milk prices. Many do not encourage the next generation to continue farming and often sell land for top dollar to developers. Except for land along the Connecticut and Merrimack rivers, the Granite State is rife with glacial till and exposed ledge – not much good farmland. The most workable soil is also most desirable for houses and shopping centers. Once it's gone, it's gone. Too bad for farming.

Our Applehurst Farm is not short on scenery. The soil is rocky, but beauty abounds. My family has always maintained open space for public enjoyment. We own Epping's two tallest hills: Red Oak and Kennard.

Vistas include landscape around our several sets of New England-style buildings, especially the farm's sweeping view from 663 feet above sea, looking down on our main farmhouse, barns, ponds, sawmill and old cider mill. The 70-acre field draping Red Oak Hill is unique for its visual vantage and recreational use. School groups and neighbors regularly enjoy the elevation for viewing stars, blue sky and rolling terrain. Churches conduct 'sunrise' services there while hunters roam the perimeter. During winter, the snow is worn thin from sledding. Mom enjoys watching people skate on the pond.

It seems there is always someone playing on 'Harvey's Hill'. When school is closed during snow storms, kids walk from all over town dragging their sleds to slide down the hill. In the old days, Mom made big pots of cocoa for us to heat on a bonfire at the bottom of the slope. Now she and Erica have installed a hot cocoa machine in The Milk Room so they can sell drinks to the visitors.

On the Fourth of July, many people watch fireworks in area towns from the crest of the hill. And Dad's airstrip! He is forever clearing just a little more of it with hopes that "Bill and Hillary" will drop in. He has pushed back the open-space borders of that 70-acre hill from forest to orchard to hay and cornfield for as long as I've been around.

"I just go back and forth, back and forth, ripping and tearing," Dad says of his perpetual land-clearing efforts with tractors, excavators and bulldozers. "And, Oh Baby! Can't I do some destruction!" he said of his most recent 'toy': his 'Little Yumbo'.

Trouble is, he sometimes gets ripping and tearing in areas where such good intentions are not appreciated. This year he began knocking down trees and enlarging a small woodland pond mere yards from a stone wall that separates the farm from land deeded to Matthew when he got 'The Crocker Place'. My niece Sharon has spent the past three years building a rustic art studio in woods owned by her parents. She milled her own timbers and did her own stone masonry, hauling everything in by wheelbarrow; one-woman, one-wheelbarrow style. So here comes her grandfather sweeping aside the stone wall and tearing down trees, closer and closer to Sharon's retreat.

I asked my niece if the excavation bothered her. She said she didn't especially appreciate having her water supply muddied and her privacy threatened.

"But we all know that if we say anything, he'll just do it more," she laughed, adding: "Besides, it gives Grampy so much pleasure!"

Niece Analesa sent this description of Grampy at age 88: "While riding my bike, I encountered Grampy who was just finishing up a day of excavating out at the gravel pit. He was standing on the tracks of the machine while he talked, which I would say is a good three feet off the ground at least, so I was wondering the whole time what would happen when he took that next step. Turns out that his procedure is to make his way along the top of the tracks to the back of the machine, grab an aluminum stepladder he keeps on top of the motor, drop the ladder along the side and then make a graceful – majestic, I might say – descent to solid ground. When picturing that, you need to remember to imagine the scent of diesel exhaust, crushed trees and a dry afternoon in April."

Other idyllic rural views include adjoining farms acquired over the years: 'The Crocker Place', where Matthew's family lives; 'Edgerlys', which is Phyllis's; Bozzie's', which Susie and my brother-in-law Tom bought from a retired Boston cop who lived out his final days in a local nursing home; and 'Chris's', which Susie also acquired. Each of the other five of us 'Harvey kids' also have homes within walking distance of the main farm. But we don't qualify as agricultural scenery, since Albert, Cindy, Beth, Erica and I each built in the woods, out of public sight. None of us live in the main 'Big House'.

Beyond the picturesque farmsteads, there are glimpses of actual farming activity. Even yahoos on souped-up ATVs can appreciate the charm of a vintage John Deere tractor driven by a hunched-over farmer as he turns soil just two furrows at a time. Or the dreamy aesthetic of a rippling sea of mature grass waving its silvery sheen in soft warm breezes. Or, during summer's ongoing harvest, the aroma of fresh-mown hay raked into wind-rows across any of our rolling tree-lined fields. Or a herd of cows contentedly lying in the pasture chewing their midday cuds – (not that THAT's a scene anyone will see at Applehurst Farm now that the cows are gone).

At present, most of the Harvey cropland is leased by the Fernalds of Nottingham: David, Barbara and their son and daughter. At least three generations of Fernalds and Harveys have grown up together and remain close friends. Fernalds have continued milking about 200 cows and need our land for hay and corn. God Bless The Fernalds! We all hope they can afford to continue farming, since our parents need the lease money.

Applehurst Farm still looks good as a working landscape, though we're not the ones working it. People still think of us as a 'family farm', not realizing our dependence on outside revenue sources.

Our first summer back in Epping (when we sold our house in Maine and began wintering in New Zealand) I was adrift as to how I should earn money. It felt like an opportunity to try farming again and work with Mom in her garden. The plan was to sell vegetables at her well-established stand in the driveway. If nothing else, it would give me a chance to catch up with my parents.

That year, I became a fixture growing and pushing produce onto whoever visited the stand. Everything seemed to sell well, except sweet corn which Susie said was 'an embarrassment'. And it was.

Come September, season's end, Mom and I sat at her kitchen table counting piles of dollar bills and tube-rolled quarters, dimes, nickels and pennies for the big bank deposit. We organized around $2,500. It pained me to set aside $400 for the jelly lady because she kept better records than we did and guaranteed herself a profit even though her product was sometimes stolen from its shelf.

Mom said I should get something for my efforts, at least for the onions I grew. I said $300 would be enough – though it meant I had worked since June for just $300 and confirmed my suspicion that growing vegetables would not provide minimum wage. When I told my youngest sibling, Erica, that I had worked all summer for $300 she scoffed: "Well! We all do it for nothing." So, I donated my earnings back to the farm for a life insurance policy we are expected to pay premiums on.

For me, 'living off the land' is a non-starter. Ironic, isn't it, when good old-fashioned agriculture just drags a farm down?

Applehurst Farm is now supported by rental property, income collected by our parents, still mostly going for real estate taxes. It's all good. Or at least it was. I mean the concept of a 'family farm' remains. It's the thought that counts. Even without cows, the public seems to appreciate Applehurst's beauty.

Hopefully, people can still enjoy looking at our scenery without farmers and livestock. Visitors might imagine what it was like to live on a farm. But, imagine as hard as they might, current generations will never really know. They CAN'T know how farming feels for my family any more than my brothers, sisters and I can know what it was like for Grampy and Grammy Peterson to convert a brickyard into pastures and hayfields. We CAN'T know what our Harvey forebears went through to clear land with oxen and plant apple trees along a rutted dirt road, impassable after blizzards and during mud season.

Apart from agri-industry, a farm isn't just a place to marvel at while driving past on the road. It's an experience – the result of a process – a family's life centered around land, gardens and animal husbandry. Trees. Bodies of water, swamps, brooks, springtime run-off and mud holes. Snow drifts and power outages. Wild animal habitat. Neighbors.

Vegetable customers. Milk customers. People who want manure and loam for their gardens. Children from the new housing developments. Horse people trotting through. Dog trainers using the pond for retrievers. Truck drivers. Neighbors and other farmers. Chickens, cows, heifers and newborn calves. Because, at the end of the day, a farming lifestyle is about people looking after one another – the natural result of caring for the agriculture.

I'll never know how my Peterson grandparents raised Mom, Alberta, Philip and Ann. From stories I've heard, I can only imagine them crawling out of bed each day at 4 to milk by hand. Then, they loaded their jugs into a Model A stake-body truck and picked up other farmers' milk from Leddy's store on Main Street, Epping. Grandparents Albert and Eunice delivered the local product to Batchelder's Dairy in Manchester (25 miles away), then picked up blocks of ice cut from Lake Massabesic on their way home.

They probably didn't make this trip EVERY day. There were so many other daily chores. But that's the point. We can never truly know what it was like without having had the personal experience. I am the only one among 16 grandchildren who remembers making the milk-ice run with Grammy and Grampy. I sat between them, jammed in the middle of the roaring solid-rubber-tired truck. Most specifically, I recall loading ice from the lake. Grampy chucked the blocks up to Grammy in the back of the truck. Then, back at the farm, I helped pack it in sawdust inside the woodshed.

I think I remember my grandparents hauling in loose hay on a wagon drawn by two horses. But Aunt Ann assures me it's just a story someone told, since even she doesn't remember work horses at Brookvale Farm. You see, we TRY to live the good life even if it's just in our heads.

Today's industrial-scale dairy operations have little to do with 'family' farms. The charm is gone. It is irrelevant. We all know that. Giving up family cows and forgetting how to milk by hand spawned generations disconnected from small-scale agriculture.

"City kids think milk comes from the store," Mom told us in 1957. Must admit I didn't see anything wrong with the concept, if that's where they bought it. As an adolescent, why would I care if kids in Manchester and Portsmouth never milked a cow or stepped in manure or saw a calf born? They probably had bottles of milk delivered to their doorsteps by local dairies like Batchelder's or Clover Leaf. Wasn't that 'The American Way'? It sounded so much easier than tending the herd and growing and harvesting their feed and selling their milk and worrying all the time about how to pay the bills. How civilized it would be to step just one slippered foot out the door to pick up a pretty bottle of processed milk!

I didn't even have slippers and had a hard time relating to TV idol Annette Funicello of Mickey Mouse Club.

Where was Bart Simpson when I needed him?

Isn't he the one who says: "Don't have a cow!"

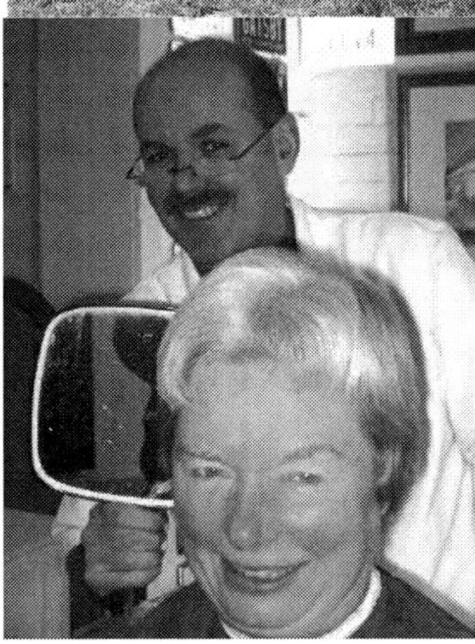

Above. The hope for the future: Nieces Christa and Abbie (l&r). Center. Bonnie (hard-working tenant at the farmhouse) arm around Cousin Olivia from Australia.

Left. My Cousin Walter Burtt (Olivia's Dad) checks to see if I have anything in my head at his popular barber shop in Ballarat, Victoria, Australia.

The way we were ...

40. SHOVE OVER!

Fitting in was always a problem at Epping Central School. Being accepted among peers is everyone's gauntlet to run, but actually – physically – fitting into a classroom was also a problem as I went through Epping's public schools: Central (elementary) and Watson Academy (the old high school).

There were an unprecedented 60 students in my first-grade class, enrolling one-third more of us than any previous class. Back then, they called us a 'bubble'; later "Baby Boomers." With this bubble promising to burst, taxpayers considered a building project: two more classrooms and toilet areas with painted masonry walls and tile floors, instead of dusty rough stone and cracked concrete, plus an office for the principal.

It took several years for this to come about.

In the meantime, until Grade 7, a dozen of my classmates were held inside a one-room school three miles away in West Epping. That kept them out of our way for a while. They were sure wild when they joined us in 1958! They swore. Not nearly as well as today's kids, but pretty good for Epping. They all came on the same bus and taught us dirty words to prove that their growth spurts had been stronger than ours. I felt like a toddler next to the curvaceous scarlet-lipped Myrna Guyette.

Elementary separation was necessary because there simply wasn't enough space at the Central School. With the basement devoted to toilets and a furnace, the brick primary school featured four classrooms on the second floor. It was supposed to be one each for Grades One through Four. Not two for First Grade and the others could shove over.

We kids knew nothing about state standards for classrooms, how many pupils were supposed to fit into square footage. But we were aware enough to realize it wasn't fair that some of us didn't have desks and had to sit on milk crates.

There was always talk of having to go 'double sessions'. Nobody wanted to go there.

It took a long time to get even the furniture right. My friend Frances was (what we now call) 'legally blind'. She knew it, but we didn't until Third Grade. That's when her special tilt-top desk arrived so she could see what she was doing.

And get this! One of my classmates, Carol Croteau, had it roughest. She grew up among a family of boys and lived in a well-worn house, quite visible in downtown Epping. Long since demolished, the two-story structure looked more like our horse barn than a house. It had asbestos-shingle siding and drafty windows. Year-round, the front door was left open and the Croteau kids didn't stray far from the crooked steps leading into their weathered wooden entry.

Carol was a kind soul and put up with a lot. Although she went the whole 12 years with me in our grade, we never had a class together because she was always in the other section ('general' or 'business') in high school. She had snarled mousey-brown hair and a plain face, plus she wore thick glasses. But, mostly, she was thought to be slow because of her hearing aid: an old-fashioned pink-plastic thing hanging off her right ear connected by wire to something in her pocket. The hearing aid squealed and seemed to make Carol's voice flat and sad. She was not 'popular' in any way. But we did graduate together. She was one of our 20.

Years later, long enough for Carol to marry and raise an eight-year-old boy, I ran into her at the grocery store (in nearby Raymond). She looked good and her voice was different, not flat or sad. She made a point of telling me about her son's ears.

"Remember how I couldn't hear in school?" she began and told me that her son was having the same trouble when he entered school (in Raymond). So she took him to an ear specialist who figured out why the boy's hearing was impaired.

"You know what it was?" she tsked.

"EAR WAX!" The doctor found that her son's inner ear had an extra twist that collected an extraordinary amount of wax.

"So, then they looked in my ears and found that I had THE SAME THING!"

She left me speechless and I've never seen her again, since our class has had no reunions. All those years in school! I was ashamed of the way we had treated her.

Did I say we've had NO class reunions? That's right. There were two 'calling-all-classes' events. But no Class of 1965 reunions. "Why?" you might ask. Or you might not. Regardless, here's my take on it: Aside from awkward relationships like mine with Carol Croteau ... and two marriage/divorce situations among classmates ... I blame it on The Vietnam War. Because of the draft, EVERY male classmate went into the military. And the few of us who went on to college (probably) protested The War. I did, at least. And, with our mere 20 drawn into such conflicting directions, the fabric of whatever we wove as classmates was stretched past the point of comfortable 'Happy Days' banter. That's the way it feels to me.

Getting back to our public education…

At the start of it, I began catching on that not all New Hampshire public schools were alike. Later, it was being advertised. An image sticks in my mind: a photo in a regional planning magazine that showed a highway sign posting miles to each town: "EPPING - 4 ... EXETER - 4" with a caption stating: "Equal Distance: YES. Equal Opportunity: NO." Photograph provided by Peter Planner. And that about summed it up for the next decade of my life.

Because there would eventually be eight of us Harvey kids and because we have homegrown parents, it is in our character to ALWAYS be loyal to our school. Each of us – from me right down to 'The Babies' Beth and Erica – (and subsequent EHS enrollees, the grandchildren: Analesa, Sharon, Danny, Chris, Jon, Johnathan, Christa, Tara and Abbie) – have defended Epping Elementary and then Epping High School until graduation and beyond. We are just too much a part of the scene to acknowledge flaws.

What did it matter if, in my class, ten of us were integrated into the next-higher grade for two years and then dropped back a year into our own class? Who cared if we had to do everything in shifts because we were too hard to keep track of as a large group? So what if we had no gymnasium and our only physical education involved fire drills and seeing how quickly we could 'duck and cover' under our desks in case Russians hit us with atomic bombs.

One year, our favorite playground equipment consisted of two freshly-cut tree stumps that we girls would line up to run at and leap from spinning our rainbow petticoats while boys stood by watching. The fun didn't last long, though. Teachers petitioned the School Board to have the stumps leveled closer to the ground and we had to resume pitching marbles across the dirt for 'Bunny in the Hole'.

Classmates who failed to thrive didn't last. Sixty of us entered the system. Twelve years later, only twenty came out the other end. Seven of my friends left to attend other schools. Whole families actually relocated to nearby towns specifically for what they claimed were better schools. Two of my classmates decided they were lesbians and ran away together. Four others didn't finish high school because they fell pregnant.

There was no sex education in school, so I didn't even know what gay Linda and Jackie had run off to be. I certainly didn't understand pregnancy. Mom had suggested "just go watch the bull" and told me: "Never let a boy put his hand on your knee." So that took care of THAT! Still, I wasn't surprised when Mary, Hope, Myrna and June were missing in September of our senior year.

Our classrooms were so crowded that 'personal space' hadn't been invented yet – let alone The Pill. We were so close that, right during class, the boys had been 'feeling up' the coincidentally later missing girls, asking them questions like: "How can you have so MUCH?" and then walking around with their little hard-ons. By the end of 12 years, I was very tired of smelling the boys' farts and hearing their nasty jokes. They made fun of my 'virgin ears'.

My first and last years in Epping schools began much the same way: Reluctantly. Rather than ride the bus, I walked to school and was late every morning of our first week back. I told everyone I hoped a car would run over me.

That was the year Susie came in as a freshman and took over. Everyone commented on her blond hair and her entourage of boys. She had clear skin while I had pimples. Among a total student population of 100 she was immediately 'popular' and was elected president of her class and a member of Student Council. As her star rose, I wished mine could shoot off and escape the gravitational pull of (bumper-sticker) "Epping - The Center of the Universe." Just before I graduated in June, Susie made fun of me in front of her Pandora-sweatered friends. A year earlier, my boyfriend had gone into the Army.

"Paul's penis was shot off in Vietnam," she taunted. "Didn't you even notice?" Just a joke. But it embarrassed me on several levels and was so typical of the way we had begun gleefully rolling over each other's feelings in favor of looking cool to outsiders.

There was no point in moaning about our education. Dad had survived Epping schools and if it was good enough for him, then it was good enough for us. He even had our same teacher, Miss Odiorne, for fourth grade! He said she had "made him the man he is today."

Mom did what she could. She actually ran for and was elected to the school board for three consecutive terms This, from a woman not keen on politics. And it wasn't as if she had spare time. She worked ALL the time. That's how important our education was to her.

She sort of apologized when I graduated.

"I knew it could be better," Mom said. "That's why I joined the school board and pushed for a new school." She said she was sorry it took so long, since the new junior high school wouldn't open until the year after I graduated. "Twelve years!" she exclaimed. "Who could've guessed it would take 12 years?!" Calling it a 'junior' high school – rather than a 'high' school – was the only way the school board could get the bond issue to pass.

The reason? Well. There were so many people opposed to Epping maintaining its own high school that, each year, building proposals failed because they couldn't get the two-thirds majority required to pass a bond issue to fund an expansion. Supporters were in the majority, but not quite majority enough.

There were two primary opponents: 1) People who didn't care about the school because they didn't have school-aged kids, and 2) People who cared about the school, but thought Epping should join a regional plan – not go it alone.

On the sidelines sat a growing number of 'new' people who didn't seem to care one way or the other. Everyone said the newcomers were from Massachusetts, often adding that they were 'poor' and had come because it was cheaper to live here than anyplace else. Judging from the handful of kids who drifted in and out of my class each year, I'd have to agree that there WAS a transient factor increasing among us. Pretty much always these kids were from nearby Massachusetts.

Later, during my college years, I worked on a survey to determine that most of the new people did not register to vote and usually lived in rental property. The survey was submitted to the selectmen as evidence that the public was losing interest in town government – since a few vested landlords (rather than many disinterested tenants) paid taxes and voted their self interests.

But not everyone was aloof. Some citizens (the ones who participated, like Dad) remained vehemently engaged in town affairs. There were no secrets. These activist types knew which side they were on as they walked into Town Meeting for just one purpose: Vote either "Yes" or "No" on building a new school. Open discussion was a formality that changed nobody's mind.

The two sides had strategies to boost their votes. When the moderator announced that the polls would remain open for one hour, monitors would rush from the hall and drive to the nearest ally's home to phone and spread the alarm. "The vote's on! We've got 50 minutes, till midnight!!" And they were OFF! Designated drivers hauled in as many consenting voters as they could find. Women arrived in bathrobes. Some of the men were drunk. And who could blame them at that hour?!

Being a loyal EHS alum, I'd like to claim that Epping behaves no worse than any other town. It would be nice if I could say that. But – after what I've seen elsewhere – I've come to believe Epping really IS more contentious than other towns. I think my hometown has more than its fair share of 'characters' who take pleasure in eroding the efforts of others. Every time I look at Epping's good-enough-for-us 'shoe box' design high school I remember the arrogance of our school board neighbor who lied to get a piece of Harvey land.

I saved some of the old 'box holder' political flyers that were distributed just before town meeting voting days. They are entertaining in a perverse way – the drawings, the claims and counterclaims about Epping schools and municipal infrastructure – including Dad's favorite: zoning and the master plan written by the planning board to outline land development.

Today, we recognize 'dysfunctional' families. Back then, we had a dysfunctional town.

It bothered me enough to take university courses with names like "Public Policy Disputes" and "Getting to Yes" and several on conflict resolution. Beyond college, I completed TWO two-year programs: "New England Rural Leadership" institute and "Community-School Volunteer Coordinator" training. Plus, I can hardly believe I spent a year driving to Cambridge one night a week for seminars at Harvard and Massachusetts Institute of Technology just to understand Epping politics!

Never did figure it out

Our parents have been to Hawaii several times. This time, it was for a Farm Bureau convention. Another time, Dad nearly died from an out-of-control infection. He snapped out of when a Polynesian minister 'laid on' his hands.

Epping Men: Cousin Dana Burtt, Tim Bruno, David, Bruce Allen, Matthew, Sam Weaver and Daniel Webster Harvey.

Sledding on lawn: Otis, Danny, Beth, Johnathan and Amos.

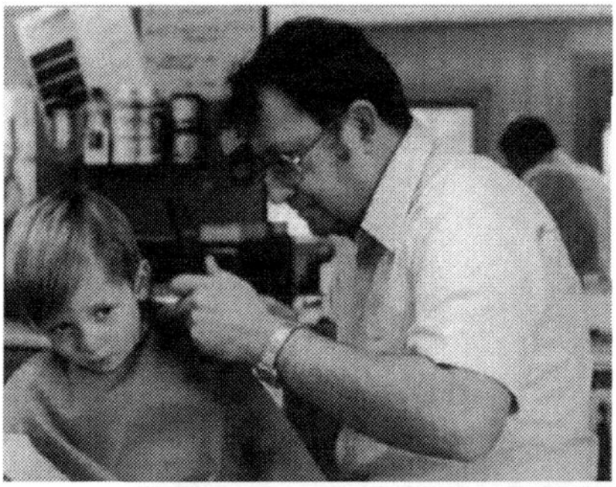

Otis's first haircut. I went to school with the barber, Armand O'Barton.

Amos, Otis, Grandpa Ted Clapp

The way we were ...

41. PROGRESS BY MASTER PLAN

While the majority of Epping citizens managed to pay their higher tax bills without complaining, larger land owners (us being the largest) began squawking because they (we) owned miles of roadway frontage. It was difficult enough trying to earn a living by milking cows without being blind-sided by an overblown tax bill, unrelated to farming. Suddenly, frontage was taxed as houselots, even though subdividing was the last thing on our minds and it was annoying to have people continually suggest that we do so.

Dad said we were "subjected to the whims of the majority." High property taxes were a real drag on the farm. Selectmen, statewide, said they had to raise assessments because the town needed more money to improve public services, police, water and sewer, solid waste disposal, municipal buildings and an occasional new fire truck.

Population was increasing faster than we could keep up. It was every town's story.

Of course, it wasn't the newcomers' fault that Epping suffered mid-20th-century growing pains. And, in fairness to Dad, it wasn't his fault either.

Uncle Eddie said it was the planning board chairman's fault for looking like Vladimir Ilyich Lenin. Ed said Epping's first planning director was the spittin' image of Lenin and it wasn't good. My uncle said he couldn't shake the similarities from his head and found Peter Planner irritating. Must admit that the man and Lenin DID look a lot alike and, once that clicked, I couldn't look at him (Mr. Planner). I closed my eyes whenever he spoke at Town Meeting, listening to just his voice.

The West Epping newcomer was one of New Hampshire's first paid lobbyists. He had an office in Concord, handy to the state Capitol, nearly an hour's distance from Epping. (By the way: May this unnamed citizen Rest In Peace! He did what he thought best.) While many felt fortunate to have such an influential professional live amongst us, others – like Dad and Uncle Eddie – wished he'd move closer to his work and get out of their hair. Their wish was NOT granted.

Peter Planner soon settled in as a resource to Seacoast Regional Planning in nearby Exeter. Before Dad and my uncle caught on, a portion of their tax money was allocated to pay the consultant and others to write ordinances to control them. Their worst nightmares were coming true. Chairman of the Epping planning board AND a regional authority! Dad and Uncle Ed reeled with the realization that Epping was destined to have zoning and subdivision regulations. The master plan would be adopted. People would have to get building permits. What could be worse?!

Well. Probably having your daughter seek Peter Planner's signature on her subdivision and building permit application. Yes. I did! And I even had it signed by Dorothy Hall, who threatened to get a restraining order on Dad because he called her "a nice round woodchuck." Dorothy was the chairman's apprentice and is now an Epping celebrity as the wicked stepmother in Meredith Hall's best-selling memoir "Without A Map." Check it out for descriptions of Epping that make mine sound like The Hallelujah Chorus.

"Peter Planner just wants Epping to be a bedroom community!" claimed Uncle Eddie. "He says he wants growth, but look where he's putting the commercial/industrial zone! Swamp and brickyard clay pits. You can't expect businesses to build in a swamp!"

As years have passed, it seems that my uncle was right (again). Mr. Planner did squeal on about "orderly residential growth" to accommodate newcomers. He was not shy about promoting Epping as a "bedroom community" for people who worked in Massachusetts. He, as our knowledgeable planning chairman, designated nearly 100 acres of clay swamp land as the only place in town where commerce and industry would be welcomed. This area is at the intersection of state Highways 125 and 101 for easy truck and automobile access. They said it was important to bypass Epping's downtown 'Historic' zone.

It worked. Nothing went into the swampy Commercial/Industrial zone for the next 30 years. The entire town was bypassed. Dad reveled in pointing this out. He said it was the zoning that scared businesses away. He pressed for deregulation and repeatedly said he intended to do whatever he wanted to do with our farm land. You name it. He threatened to do it: regional landfill, expanded junkyard, stone crushing to flatten Kennard Hill, trailer parks, hazardous waste recycling, and municipal sludge disposal. He didn't seem to have a high-tone thought in his head.

With people like Dad taking pot shots at zoning and other planning board initiatives, businesses didn't want to take a chance on Epping. We continued near the bottom of state rankings for per capita income and spending on schools. If Dad's goal was to maintain the status quo by throwing wrenches into the works, I say mission accomplished!

We got through it.

I don't know how this era will come out in history books, but it seems to me that very little improved for Epping until we got into the 1980s when newcomers outnumbered natives and people from Massachusetts were no longer scapegoats.

Unfortunately, by the time things calmed down, Epping had been eclipsed. Less feisty neighboring towns of Exeter, Stratham, Plaistow and Raymond were considered more progressive and stable, more suitable for business. They were the first to strip-zone and fill their commercial and industrial areas with supermarkets, manufacturing and offices. Epping residents had to drive out of town to buy groceries, clothes, building supplies or to see a movie.

'Progress' eluded Epping until just recently, when the big-money interests finally nudged aside environmental concerns and moved into the swamp. Wal-Mart, Lowe's, McDonald's, Burger King, Starbucks and dozens of other stores now occupy major and minor shopping malls in the wetland industrial park. Target, Market Basket and The Home Depot are next on the horizon.

It strikes me funny that, all of a sudden, Epping is looking pretty good compared with the towns that sprinted ahead of us 30 years ago. Even our residential areas seem superior, having side-stepped the push for low-income housing by being too disorganized to accept federal money. Our one shoddy apartment complex, named "Watergate" because it was built in a swamp behind the Central School while Nixon was being impeached, was demolished recently and replaced by a modern Tupperware-looking multi-unit thing floating on an acre of asphalt that should last a while.

Neighboring towns' once-new malls have deteriorated and their downtown districts lack vitality after all the stores were moved to peripheral shopping areas making everyone drive to get there. Groceries are no longer available in the center of either Exeter or Raymond. When I was a kid, I remember Exeter, in particular, having everything we needed right downtown. Now, it's just a bunch of boutique shops catering to Phillips Exeter Academy.

Of course, Epping's downtown hardly exists compared with Exeter's. Still, it has also changed dramatically. In my lifetime, four grocery stores have closed along Main Street – The A&P, Mike's, Fecteau's, and Leddy's pharmacy, soda fountain and grocery. All gone. If it weren't for commuter traffic dodging lights on Route 125 by short-cutting through town, central Epping would be near dead.

But when it comes to shopping, Epping malls are bustling, the newest and best in the area. So let 'em come to Epping for a change! We have 'anchor' stores, FOUR Chinese restaurants and several furniture outlets.

When someone writes this up, my parents may be recognized as worthy guardians of the ground they feel obliged to protect. Standing in the way may be an effective way to stop progress from running over you, at least until the development machine is tuned for better results.

It's all for the future generations: Christa, Amos, Analesa, (pregnant) Melanie and Otis.

Mom and Dad
1997

The Harvey Girls:
Aunts Aggie Wilcos and
Edna Woodward

That's Aunt Aggie's
hand on Uncle
Casimer Wilcos's
shoulder – not Ken
Buxton's.

"I'd have to say it appears that someone had a hard landing." – Phyllis Harvey

Part Four – To Protect The Innocent

PART FIVE

Farming Flame Gutters And Dies

The way we are ...

42. UNCLES PASSING IN THE NIGHT

Note: There are several "almost true" things in this section: Cindy did train oxen – two at a time with Uncle Philip's help. But t'was I who raised the freemartin named 'Pepper'. The odd animal was devoted to me. The sexually-confused cow was, rather a dolt and quite bullish– not nearly as interesting as the composite 'Jerry Lee'). Cindy's oxen and Sadie Lee were Holsteins, not Ayrshires. Pepper was fluffy white with black speckles.

We lost two uncles while David and I were back in Epping last summer. Uncle Bill Woodward on my father's side, married to Aunt Edna. And Uncle Philip Peterson, my mother's brother, married to Aunt Virginia. This never happened before. Two gone in one summer.

Uncle Bill could always be counted on for a big smile and strong handshake. Even when I was small, my uncle pumped my hand and declared: "It's SO very good to see you!" This continued into my student years at the University of New Hampshire when I dropped by my aunt and uncle's for lunch. Woodwards' farm abutted the university along Main Street. Bill's father had been a professor of forestry and Aunt Edna worked on campus as a Registered Nurse.

Since my uncle and, later, my cousins went straight through Durham's school system, they knew everyone in town. Within the tight-knit college community, I enjoyed a second-hand hometown advantage of background information on my professors, their families, and campus staff I worked with through 'Work-Study'. Cousins Margaret and Jean attended UNH when I did, graduating in Library Science and Occupational Therapy. They are both very smart, as is their younger brother, Cousin Harvey, now retired after a career in the timber industry.

Uncle Bill raised sheep and milked a few cows, but mostly he operated a sawmill and worked in the woods. At his funeral, there was a terrific picture of him as a young man with his horses, so strong, yet gentle and willing – uncle and horses alike. Bill was like that, a tall strong man with the natural grace of a well-read satisfied husband and father.

That's why it was so sad at the end when my uncle lost control and was miserably in and out of intensive care and rehabilitation centers. During his three-month decline, Uncle Bill returned to Aunt Edna and the farm several times only to be rushed back to the hospital and fitted with tubes.

David and I visited him in the hospital. Took my father with us once. It was the only time I recall him being at a loss for words – Dad, not Uncle Bill, who was moaning and mumbling constantly, rolling back and forth against restraints on an emergency room bed. Before we arrived, he had rocked his way toward the bottom of the bed with pillows knocked to the floor and sheets shifted to the side. One of my uncle's long muscular legs hung out through the cold pipes intended to cage him. He repeatedly extended the leg as if searching for something to push against.

I grabbed hold of his flailing foot because it seemed like the most obvious way to be supportive. He promptly shoved himself back toward the top of the bed and calmed down. I'll never forget the one clear thing Uncle Bill said and did that night. He smiled his big toothy grin and pulled up his shorty nightshirt to flash us, warmly declaring: "All dignity is GONE!" Wouldn't you have thought they'd leave him his underpants? ... which I guessed was his message.

Dad was speechless and declined my offer to caress his brother-in-law's foot while I went to find a nurse. I could only assume he was hoping he wouldn't go that way himself. But he never said.

Bill's passing was, of course, especially hard on Aunt Edna and our cousins Margaret, Jean and Harvey, who have always been an important part our lives. They provided loving round-the-clock care for their father right up to the end. With Margaret and Harv living nearby, Jean has now moved home to be with their mother. This first winter without Bill stoking the fires and reading beside the cookstove will be difficult.

Bill's wake and funeral brought everyone together to honor an honest man who had done his level best to live and love well. Friends who I hadn't seen since college stood up to praise Bill as a "father figure" and "role model" who provided guidance lacking in their own families. I didn't realize my long-ago peers had continued such steadfast contact with my relatives.

We basked in the warm awareness of our shared family, which was sweet enough without our cousins' a capella singing. Lynn and Eileen arrived from Connecticut to deliver a perfectly-pitched salute from the rear of the assembly hall. Then Jean sang a solo hymn from the podium.

We were wiped out, drained and silent in our memories and private thoughts. I sat in awe of my cousins' brave talent, their voices so strong as they harmonized and soared to meet the dreadful challenge of losing their fathers. (Lynn and Eileen's dad, "Uncle Coz," Aggie's husband, had died two years ago and was buried on the farm in Epping.)

I couldn't help remembering another New Hampshire farmer's funeral that I had attended just a month before Uncle Bill's passing. A tragic loss when hydraulics let loose on a gang of plow disks and crushed Arthur Scruton in Rochester. He had been an award-winning dairyman on the family farm just half an hour from ours. Arthur was younger than me, but I remembered him clearly from our days as 4-Hers showing and judging cows throughout the state.

The two memorial services dredged up subliminal awareness of shared farm life and respect for our undeniable extended families through blood, agriculture, and friendship. Both passings were steeped in a greater power. Similar to Uncle Bill's service, the Scruton Family and assembled mourners shared strength, rather than weakness. Arthur's older brother, John, conducted the funeral with great sensitivity and the ease of an experienced pastor (which he is). Then, same as when my cousins sang, I wilted with admiration as Art's younger brother Danny delivered a tenor tribute.

How do these people tap such depths during days of upheaval and sadness? What is important to remember about these funeral experiences? Certainly that our family farms are being passed to the next generation, ready or not, either by disease or disaster.

What will happen to Arthur Scruton's cows? What will happen to Uncle Bill's sawmill? What will happen to our own fields and forests, the houses and barns, the dormant cider- and saw-mills, the airplane hanger, the junk yard?

Will the cows ever come home?

It's not lookin' good!

Uncle Bill Woodward

The way we are ...

43. AMAZING GRACE

Last summer while we were back in New Hampshire, my Uncle Philip came storming out to our house angry as a wet hornet because some "bastards" had cut down his favorite cedar tree on the Epping-Fremont town line.

"That tree didn't do nuthin' to deserve that!" He was steaming, madder than a man with a heart condition should get. I sat him down and gave him a glass of water. He said the Epping road crew had sawed and hauled away a tree beside the road, a tree he'd been admiring all his life.

"It was a beautiful tree! I checked it every morning. Just lookin' at it out the kitchen window made me feel good." He said there was "no way in Hell" it was hurting the road, and the maintenance workers had no business cutting it down while he was in Exeter for a doctor's appointment. He thought the Town might've hired my brother Matthew to haul it to the stump dump that sits between our property and Nottingham Square Road.

I assured him that Matthew would know better than to touch the cedar.

"He would've told you if he got wind of anything like that, I'm sure!" And I WAS sure. Anyone who knows our Uncle Philip watches out for him and appreciates that he is keenly aware of anything that moves on "the homeplace," Brookvale Farm's 200 acres.

"Why didn't Scott and Brenda stop them?" I wanted to know, since the tree stood opposite my cousins' driveway. Philip said they hadn't seen a thing and must not have been around, which was a good enough excuse, since their driveway is nearly half-a-mile long and they live back in the woods.

"Road crews can be tricky like that," I said. "Remember when they tarred Red Oak Hill the week my parents went to Nova Scotia?" (50 years ago).

"The bastards!"

Part Five – Farming Flame Gutters & Dies

Next time I heard any more about the cedar tree was when I saw pieces of it lashed together to form handrails into Uncle Philip's house, the large red Colonial where he and his three sisters, my mother, Alberta and Ann were born and raised. Philip was laid out in a coffin in the back room, always an unused dining room, where I had last seen my grandfather, Albert Peterson, similarly displayed following his death on New Years Day, 1955.

Philip had died unexpectedly "in his sleep," as they say, this past summer just two months ago. "He just laid there with his eyes open just a slit and his mouth hanging a little as if he was sleeping," Aunt Ginny said. "I told him: 'Philip! It's time to get up.' But he didn't."

Cousin Dana stood by the farmhouse door playing his mandolin during late-afternoon visiting hours, as over 700 mourners filed through to pay their respects. Yes. That's SEVEN HUNDRED mourners who realized that Philip represented something special about farming with a sense of humor.

My uncle's grandchildren and my cousins put together a first-rate memorial. They built a plain pine coffin lined with burlap, stuffed with hay and accessorized with braided bale-string. They placed pliers in Philip's hand. 'The Brewitt' did a nice job with the corpse, but my uncle still looked dead.

In the barn, videos played showing Philip on the farm. One segment made me laugh, then cry. It was Philip encouraging his three-year-old grandson to "Get in there! That's it, Lester! Get in there!" while the little boy wandered among a dozen unrestrained hogs. The pigs were as tall as Lester yet gently allowed him to float among them as free as a soap bubble. Watching that movie and listening to my uncle's warm chuckle, I knew Philip had truly loved his wife (Virginia), three children (Albert, Brenda, Linda), and six grandsons. I knew, too, that he had loved me and enjoyed watching me grow up. My siblings and cousins – Eric, Peter, Walter, Fred and Dana – had similarly benefited from Uncle Philip's steadfast support.

Dana, my tall quiet wide-shouldered mandolin-playing cousin wore a tan Carhartt work jacket. He looked so much like his father, my Uncle Eddie, in his younger years! I was surprised that Dana was a musician and mentioned it to Aunt Ann, his mother.

"We didn't know either," she whispered. "We're trying to get closer so we can hear better." Uncle Eddie stood off to the side, back from her. It's too bad, I thought, that they don't get along better. They wouldn't have to be so coy.

There was a family "incident" five years ago, where Uncle Eddie became unreasonable over a 'Peace' symbol that Dana's wife, Luanne, displayed on the side of the barn. She considered it a decoration. But Uncle Eddie didn't think it was right at all and pretty much freaked out when he saw it.

The offensive ornament was crafted from that plastic Christmas light-rope stuff that you can tack up or drape and you don't have to worry about moisture shorting out the bulbs. I think Luanne had formed it from a Hula-hoop.

My aunt and uncle's daughter-in-law thought she was doing a splendid job of getting festive for a big Christmas. She is a professional interior decorator and had spent pretty much all year re-doing the house and grounds for the reunion. Even her brother-in-law's (my Cousin Walter's) family was coming from Ballarat, Australia. Luanne had never met them, since Walter had not returned to Epping for many years.

Cousin Dana and Luanne live in the house where he and his brothers (Peter, Walter and Fred) grew up. They moved in about ten years ago, when Aunt Ann and Uncle Eddie retired and moved to Nova Scotia.

During his prime, my uncle was one of New Hampshire's finest bulldozer operators. He engineered and installed several classy golf courses and was known for his attention to detail and ability to work all day without taking a leak. He even engineered his own 675-acre private park in Northwood and sold it to the State. Now known as Northwood Meadows State Park, it features an Ed-made lake at the headwaters of the Lamprey River. Hiking and picnic areas throughout "Betty Meadows" are now enjoyed by the public.

Uncle Eddie is that good as an earth-moving artist.

Part Five – Farming Flame Gutters & Dies

His and Aunt Ann's retirement feels like a self-imposed exile to Canada, back to land of their ancestors – my ancestors, too – fairly mutual ancestors, Ann and Eddie's, since their (and our) Canadian grandmothers were from two separate (I hope) families with surname: 'Cole'.

My aunt and uncle say they want no more to do with Epping now that their sons are grown, married and one deceased. They seem to have come to terms with Canadian 'socialism' by crossing the border for American medical and dental appointments. I think it's a 15-hour drive, which they often do without stopping. Sometimes they take a ferry from Yarmouth to Bangor. Ann does all the driving now and sometimes comes by herself. Last summer, she made a special unannounced trip down just to give her Australian granddaughter a "talking to." Spent a night at my folks and headed back the next morning. Not bad for a woman who used to become so homesick for the farm that she'd haul my cousins up to Nova Scotia and bring them all straight back because she missed home.

So, here came the Burtt Family to celebrate Christmas in the house they had built during the 1960s. And there stood the weathered-gray barn, measuring 40-by-60 feet, which Grampy Peterson had disassembled and moved 15 miles from Hampton in 1951. The home place was not without personal attachment.

My aunt and uncle were first to arrive from Nova Scotia. And Uncle Ed wouldn't get out of the car. Parked in the driveway, he told Ann they weren't staying with that "communist symbol of the devil" on the old barn. Aunt Ann was exhausted and began sobbing. When Luanne and Dana came out to greet them, things turned ugly. Luanne was tired, too, and didn't appreciate being criticized for her efforts to beautify the place.

"I told him it was a Peace symbol, not anything to do with the devil or communism," Luanne said afterward. "But my father-in-law was raging and told Ann to drive him back to Canada." Dana was caught between his wife and his parents, tending to side with Luanne who said she'd call the Community Church minister to get it straight about the Christmas decorations.

Uncle Eddie said he didn't care what the minister said because he wasn't even a Baptist. Dana offered to get a ladder and take the lights down. His father told him to go apologize to all the neighbors who had been driving past, having to be offended by the Symbol of Satan. This didn't go over well with Dana who raised his usually-soft voice to tell his father he was being unreasonable and pigheaded. Ed yelled back while Ann and Luanne stood by crying. Before long, Dana was crying too, he was so angry. He got a ladder, climbed it in a fury, tore the lights from the side of the barn, and tossed them to the ground.

"There!" he told his old man. "Are you happy NOW?! Merry damned Christmas!!"

Cousin Walter hadn't been back from Australia for 15 years and said, later, he couldn't understand why everyone was so quiet and why the parents were staying at Aunt Louise and Uncle Dan's. Nobody told him what had happened until Ann and Ed left for Nova Scotia the day after Christmas.

From that point on, my aunt and uncle have stayed with my parents whenever they return to Epping. The sisters (Ann and Louise) enjoy each other's company and the men share a lot of history and attitude. Luanne, who once preserved Ed's office untouched for his use while visiting, loaded all of his files into cases and stacked them in the basement bunker. She threw out his desk and replaced it with lounge furniture to make a sweet little den.

So, there was Dana playing "Amazing Grace" for our Uncle Philip's wake and his parents keeping their distance, trying to hear.

"Pretty good, isn't he?" I said.

"Can't hear him. He plays too soft," Ed said.

I thought: Well! Strained relations make for strained listening.

The next day, the funeral was on television: Channel 9. Aunt Ann phoned to tell us Quick! It's coming on at seven. I had to chuckle at all the close-ups of Dana and his mandolin, providing a nice LOUD musical backdrop for the feature. His parents were watching him be a TV star, if not a Peace symbol.

"Amazing Grace!" Indeed.

Cousins Christa, Abbie, Tara

Uncle Philip Peterson's Wake. Seven hundred people signed the visitors' book that night.

Uncle Philip loved his pigs and grandchildren.

Philip Peterson and Walter Moulton at Brookvale Farm.

The way we were ...

44. PHILIP SAVES THE DAY

From inside the cat house, I hear Grampy Peterson whistling as he stacks wood against its back wall. Ker-THUNK—ker-THUNK! The hunks of split oak and maple fall into place. No doubt, he knows I'm in here with the kittens. He often says it's a good place for me, inside my favorite hiding place out of harm's way. Grammy is working in the vegetable garden.

Just another day at Brookvale Farm in Fremont, where my mother was born, seven miles from Applehurst Farm on Red Oak Hill in Epping.

I always felt I was Albert and Eunice Peterson's favorite because I am the oldest grandkid, the one who spent the most time at their farm before Grampy died in 1955. Few of my cousins and siblings ever saw our Swedish grandfather. Mom willingly let me stay with her parents for as long as they'd have me. It was a privilege. Grammy gave me my own bedroom upstairs where I could eavesdrop on them through a register above the kitchen. I never loved my grandfather more than when he persuaded Grammy that I "must be very tired ... too tired to go to school," if I didn't respond to her yelling for me to get up. Sometimes it worked; sometimes it didn't. Either way, Grammy scolded me for "playing my little game" to skip school.

I lived with Grammy and Grampy while Mom took Susie, Matthew and Phyllis to Florida for a month during the first two winters of my public school education, until my grandfather became ill. Mom said everyone except Dad and I had to go to warmer weather because Susie couldn't stand breathing cold air, though I never noticed her having a problem with it. Dad had to milk the cows; I had to go to school.

Truth be known, it was Mom, more than anyone else, who wanted to escape New Hampshire's frigid winters. She finagled her way to Florida most winters, even if it was for just a couple weeks. Once, she disappeared to help her sister, my Aunt Ann, elope with Uncle Eddie. I remember Grammy taking the phone call to learn that her baby had married. It made her cry for joy (I'm sure). Another time, Mom had to visit a sick friend. And, of course, there was Susie's breathing problem. Twice, she left us to drive down with a neighbor, Mrs. Head, because Mr. Head had died.

Dad and we maintained pretty well during her well-deserved midwinter vacations. Everyone said it was good for her to get away. Over time, even with good reasons to escape, Mom quit the routine. I don't recall it being a big deal. She just stopped complaining about cold winters and didn't go to Florida anymore.

For awhile there, being dumped at my grandparents', I got to know Mom's birthplace pretty well.

The cat house was originally built for "Shep" the dog, but he never used it because there were plenty of better places for him. The fat border collie enjoyed two blankets in the barn and one in the woodshed. Besides, the cat house smelled too much like sour milk and cats – for a dog. It was a crawl-through tunnel room, open to the outdoors on one end and to the woodshed on the other. Basically, it was just another use-specific enclosure inside the sprawling barn. Unlike the Harvey side of my family, Grammy and Grampy Peterson left NOTHING outdoors. Everything was undercover – equipment, tools, animals ... us. Just the vegetable garden and clothesline were, of necessity, exposed to the elements.

My grandparents' house was attached to their outhouse and woodshed, which were attached to the cow barn, which provided shelter for the cows, horses and pigs, plus hay storage and workshop space. Machinery and several hundred chickens were housed nearby in a shed on the other side of Martin Road, which passes within 25 feet of the house.

Grammy and Grampy's farm is still beautiful, quaint even, with its spacious white Gambrel barn and red Colonial-style house, the connected architecture and mature maples on the lawn. It is different from the Harvey farm in many ways. Much more tidy without a lot of equipment sitting around – tractors, balers, manure spreaders, trucks, harrows – no junkyard 'Up Back'. Most obvious, though, the terrain is different.

Brookvale Farm is open and flat. The soil is heavy with clay and not well drained. Two lazy brooks twist through the fields, sustaining swampy areas inside each bend.

Unlike my Harvey-Ladd ancestors in Epping, who forced agriculture onto the top of a forested rock-strewn hill, my Peterson-Cole grandparents' farm had already been cleared for them during the 19th century. Right there where the farm is today, four brickyards supplied kiln-fired blue clay materials for major buildings still standing throughout New Hampshire and northshore Massachusetts, and in India, too, since many tons of brick were exported as ballast for sailing ships.

Meandering streams and two large ponds are all that remain of the clay excavation. While pretty to look at, the brooks make it difficult to traverse the property. You can't just walk out across the field and expect to reach your destination. The streams are deeper than one might expect as they continually shift channels and slice themselves through the tufted grassland. It's not impossible to jump across the brooks, but many of us have ended up shivering and soaked to the waist trying.

My grandparents met in Norwood, Massachusetts, where Grampy grew up as a middle child among seven brothers and two sisters. Grammy giggled like a school girl saying it was "love at first sight" when Grampy invited her to ride along and deliver bottles of milk to houses from his dad's horse-drawn wagon.

Grammy (Eunice Martin Cole, born 10/30/1902) and Grampy (William Albert Peterson, born 1/19/1895) married in 1924 and moved in with her parents (Henry and Caroline Cole) to work on the Fremont farm. She said she always wore pants and worked with Grampy "just like a man" (and there was no reason the rest of us couldn't do the same). Around 1936, my grandparents purchased the 200-acres of brickyard that the Coles' had been leasing. Respecting the waterways, they named the property Brookvale Farm. It is located along both sides of Martin Road.

My mother, Louise, is the oldest of four Petersons, all born right there in the farm house:

Carolyn Louise, born 7/9/25; Eva Alberta, born 6/22/27; Philip Edward, born 4/22/29; Marjorie Ann, born 9/11/35.

Brookvale Farm's 'cat house' was too small for adults, but roomy enough for kids and cats. The primary adult-filtering feature of this 'house' was its sparse head room – as in NONE. It had two doors and two rooms, but was just a crawl-through, always dark and warm, no matter how bright or cold it was outside. Grampy made its doors from burlap grain sacks, doubled over, meeting in the middle for easy push-through access.

My grandfather was clever that way with his automatic-closing doors using horseshoe weights hanging from ropes, and niceties like strategically-placed shelves for our sticks and a pulley on the clothesline so Grammy could reach both ends. He built the large gambrel barn himself, designing a major slice off the west corner to accommodate the daily freight train. Among the barn's amenities was a long-drop toilet that Grampy installed in a little booth behind the cows. It was painted yellow and had a magazine rack nailed to the wall. Grampy called it his 'study'. He was very particular about whitewashing the barn walls and spreading the floors with powdered limestone scattered with sawdust. On Sundays, he washed all of the cows' tails and fluffed their twitches with a comb.

Once, when no one was around, I printed my name very carefully across the barn floor: CAROL HARVEY, etched into the lime and sawdust with a hoe. Uncle Philip was SO sure Grammy had done it!

Grampy told him not to be silly. "Of course Carol wrote that."

"It's too good," countered my uncle. "Too adult. Look at the A's and the O and E. Momma wrote it!"

"Why on earth would Momma write Carol's name?" asked Grampy.

"I dunno," Philip said. "Because she likes her?"

I was a winner that day! I could print like an adult and Grammy liked me.

When I wasn't hiding in the cat house, Grammy made sure I helped with the various farm chores. She always had something that needed doing: hoe manure from behind cows, sweep the mangers, fill the grain wagon, remove wet hay and "slobber" from the drinking cups, fluff the hay, dribble it with molasses, push the feed closer so they could reach it, mix grain and swill for the pigs, haul milk to the milk room, brush the cows, tote that barge, lift that bale ... and then there was the lawn to mow and weeds to pull from the garden, eggs to collect from the hens ... heaps of assignments.

Sometimes I helped bring the cows in from the pasture, riding behind Uncle Philip on his Morgan mare 'Sprite'. She was a clever horse and responded well to the chunks of carrots and cubed sugar that my uncle always carried in his pockets. He had taught her to 'neck rein' using just a halter and rope for direction. A tawny chestnut with cinnamon and blond mane and tail, Sprite was the most wonderful horse in my world. She was eager to please. Whenever Philip called her from the gate and she'd perk up with a whinny and come trotting over to pick us up and go get the cows. She wore a halter. Philip carried a rope for reins. With a few carrots here and a pat there, we were away riding bareback. Most of the time, Sprite just walked. But if Philip clucked his tongue and said: "Go get her!" the mare made a trotting transition to canter toward the wandering cow. When he clucked, I hugged tight to my uncle's waist so I wouldn't fall off.

One morning, there was trouble. I was flinging 'scratch feed' to the chickens when I heard Grampy yell for Philip.

"Sleeper's bull is loose!" he called. "The bahstid's botherin' the cows. Hop on Sprite, will ya, and get Sleeper over here. Quick!"

My uncle was up and away as fast as Sprite could carry them. Grampy half-filled a tin bucket with grain and said I could go with him to deal with the cows. He handed me my cow-thumping stick.

"Shouldn't we tell Grammy?" I asked.

"No! You keep her outta this," he replied.

The action was way down in the last pasture. Grampy was rushing, so I had to jog along behind to match his long determined stride. My eight-year-old throat was dry in anticipation. Grampy told me I would stand ready to open and close the gate to the 'race' (the fenced walkway that defined the western property line, running along the railway bed). This straight built-up cow path accessed far ends of the pastures. It was the easiest track to follow, since it bridged the brooks.

"Now, don't you go gettin' in the way, ya hear?" he said, warning me one more time about how dangerous bulls could be. "Brutus won't be happy about this!" Grampy wondered aloud how he would separate his Ayrshire bull 'Brutus' and the cows from the intruder. "Can't trust ANY of them! Size doesn't matter. Darn stinkin' bulls," he muttered. I wondered if bulls were an exception to his rule that, left alone, an animal will never harm a child.

My grandfather would be dead within a year, but that day, he was a gladiator set to battle bulls and claim his bodacious bovines. He was my silver-haired hero with a sweet Swedish accent and freckles. I would've followed him anywhere. His faded-blue short-sleeved shirt stuck to his broad back where sweat revealed a skinny-strap undershirt. His sunburned forearms and weather-worn hands pumped and swung us through the hot summer air to close the half-mile distance between us and the cows.

At the gate, we stopped to catch our breath. No doubt, Grampy was deciding what to do next. Brutus and the neighbor's black jersey bull were posturing with broadside stances to impress each other with their size. They bellowed threats and heaved dirt into the air, each pawing with one front hoof. Every once in a horrible while, they charged straight on and cracked their skulls together, pushing in a standstill, each measuring the other's strength.

The herd of about twenty Ayrshires stood at attention, tense with tails held tight against their rear udders. They did not respond when Grampy shook his bucket of grain calling: "So Boss! Come, Bossy, come!" The cows were focused on the fighting bulls – frightened, but curious. Grampy said he was going out there and I should stay back to open the gate if he managed to head the cows over this way.

As he walked into the pasture, Grampy whistled to the cows and repeated: "So, Bossy, come!"

The bulls became more vocal as they roared and snorted, pushing each other along the nearby brook. Brutus was older and had a definite size advantage over Sleeper's bull, but the younger animal was aggressive and determined to mingle with the ladies. Back and forth they went with their heads firmly butted together. Neither had horns, so that was good. Plus they wore leather halters on their heads and brass rings in their noses.

All of a sudden, there came Uncle Philip and Sprite galloping toward us like the Range Rider and Rawhide on fire. Philip whooped all the way from Martin Road, his red-plaid shirt blowing out behind him. He hollered: "Sleeper's coming with a wagon to get the bull!"

And, just as he yelled, so did one of the bulls. The smaller black one uttered a startling roar as it tipped over backwards into the brook, hooves flailing in the air as he struggled to right himself in the narrow channel. Nobody actually saw it happen because all eyes were on Philip. We could only assume that Brutus had mustered a mighty shove to end the bull fight.

It was too much for the cows. They decided flight was the way to go. Crowding nervously for a second, twenty Ayrshires jumped their liver-colored hides into the air, turned like a puff of smoke, and headed straight for ME.

Grampy was knocked over by the stampeding mob, but managed to shout: "Open the gate! Open the gate!" So I did – just barely in time to let them shove it and me aside in their panic to escape. The runaway herd was strangely silent. No bellows or snorts – just 20 straight-out tails and the rattle-your-bones clatter of eighty cloven hoofs along the hard-packed race all the way to the barn where we found them standing, each in its own stall, as if nothing had happened.

'Shep' and me at Grammy & Grampy Petersons' outside 'Philip's Little House'

My brother Albert.

The way we are ...

45. PHILIP'S FUNERAL

The Sisters' Refrain ... (please)

The fight was actually true.
It turned Erica black and blue.
In Phyllis's room
She was whacked with a broom
And the police didn't know what to do.

Last summer, when Uncle Philip came searching for the cedar tree, he asked me how Phyllis was doing with the cows.

"Probably everyone is mad at me for getting her into this mess!" he said emphatically, dropping his arms in one flap to slap his hands to his sides, same as he has always done, as did my grandfather before him.

"It's a damn shame!" he declared. Fwap! His hands slapped onto his bib overalls.

He said he'd been hearing complaints from The Sisters (Susie, Erica and Cindy) about my other sisters' (Phyllis and Beth's) farming practices. He had heard that they didn't plant the corn in straight rows and that they dumped big chunks of concrete in the fields, out the back of the manure spreader which was always broken and that they had ruined one of the tractors by installing transmission parts backwards. And, of course, they weren't making enough money to cover expenses, so our father kept paying for the repairs and diesel fuel and who knew what all. Susie had told him that the cows were costing the farm $100 a week.

Uncle Philip said he wasn't surprised that they weren't making enough money to cover taxes on the barns. "None of us ever made money," he clucked. "But is it true that she's being EVICTED?" I said it seemed to be the case, and what did he think? He said he didn't know what to think ... except that if HE didn't give Phyllis his herd, then maybe none of the trouble would be happening today.

"But what was I supposed to do?" my uncle asked. "I couldn't keep 'em with my legs being operated on and I didn't have the heart to kill them." Phyllis had offered to take the herd while he recuperated, but ended up keeping them. "I didn't want the darned things back." He grimaced and flapped his arms.

I told him that my sisters were perfectly capable of fighting with each other without his involvement. "None of this is your fault," I assured him. "Did you hear about them calling the police because Beth thrashed Erica with a broom while Cindy wrestled the boyfriend and Phyllis watched and didn't know what to do?" He said, yes, that was one of the things he had heard and that's why he was worried about the situation at our farm.

"Well. Nothing came of that," I said as if it mattered about the police. "Because Erica and Cindy went into Phyllis's house and started yelling at her for putting a fence back across the Night Pasture and Beth and the boyfriend .... Well! I guess it was pretty bad. I saw Erica's bruises. The police took pictures of them. But the boyfriend claimed he was a victim, too, because Cindy had grabbed him and that was assault. And it DID happen inside Phyllis's house where they weren't welcome. So that was like trespassing and one complaint canceled the other and nobody got charged with anything."

"Heck of a thing!" Philip snorted and flapped his arms. Fwap! "To have it come to this!"

"Disgraceful, really," I agreed.

According to Uncle Philip, it was 13 years ago when his knees gave out. At the time, he was milking a couple dozen cows, the same herd that had been on Brookvale Farm since 1942, when my grandparents raised calves to replace milkers that the State tested and destroyed for brucellosis.

All the Petersons called it 'Bangs' disease. Destruction of the animals was understandably traumatic for everyone involved, especially the farmers and veterinarians. As a kid at my grandparents' farm, I remember the word 'Bangs' spoken in solemn hushed voices. Clearly, it was a sad sad day when the herd was hauled off for slaughter. Sixty years later, it remains a family tragedy. Recently, I told a joke about cows being angry because they had Mad Cow Disease. Aunt Ann practically ran out of the room, it upset her so, reminding me that Peterson-Coles never forget.

When Philip needed surgery in 1994, he said the cows HAD to go.

"It's the end of the cows," Philip had said. "Brenda doesn't want them. She's too busy, and it's just too much for me anymore." (Brenda is the older of his two daughters. Linda is the other one. Son: Albert.) Brenda had milked the cows for a while.

That's when Phyllis offered to take them off his hands, expecting to pay Philip something monthly for as long as she milked them. Our uncle told her not to be silly. They'd be costing her more money than making it. There would be nothing extra to be frittering away on paying him.

Phyllis owns her farm, 'The Edgerly Place', given to her by our parents. The estate includes an old Cape and a two-story barn with a full cellar wide-open along one side at lower ground-level to provide space for manure storage and livestock shelter. It sits on 170 acres of woodland and fields that extend out to where David, the boys, and I live. Phyllis's farm is beautiful and – with the benefit of 20:20 hindsight – all of this being evicted trouble could have been avoided if she had taken Uncle Philip's cows to her place, instead of to the already-existing dairy facilities at the adjoining main farm. BUT, at the time – when Philip needed to get rid of the herd – Phyllis couldn't afford to build her own milking facility: a parlor with pipe line, bulk milk tank for refrigeration, loose-housing feedlot, and all the rest of it.

So (some would say) she made the mistake of reactivating the dairy at the farm. It wasn't easy putting the equipment back together, since it had been dismantled following the federal Buy-Out. Phyllis had to find and retrieve essential parts, like a compressor and milking machine cups and claws. She bought a new bulk refrigeration tank.

Back then, when the cows returned, everyone seemed happy about reactivating the abandoned plant.

Our sister Beth and brother Albert pitched in and helped with the project. Mom seemed especially pleased to have the Peterson herd saved. She and Beth spent many hours helping Phyllis train Philip's cows to enter the parlor for milking. They led each one in by halter, coaxing with a rattling bucket of grain, sometimes pushing from behind, until the Petersons' Brookvale herd adjusted to the Harvey routine on Red Oak Hill.

Phyllis worked out a milk-sales agreement with the big hauler 'Agri-Mark' who had continued picking up from the Allens who also live on the hill. Milk from Philip's small herd barely covered the bottom of the bulk milk tank that had accommodated 100 cows just seven years prior. The tanker picked up after four milkings (every-other day). Depth of the first milking was so shallow that Phyllis asked Beth to weld an agitator extension so the paddles could reach the milk.

Uncle Philip did not make a miraculous recovery. He hobbled around for months following surgery. My mother said he "couldn't afford" to deal with the cows again. The assumption was that he shouldn't subject himself to the bumping and wrangling that goes with managing cattle. He couldn't "afford" to damage his knees. But she probably actually meant it. Her brother could not AFFORD to take the cows back because it costs too much to milk as few as 20 cows.

Months passed, then a year. And, in the end, Philip told Phyllis to "just keep them." They were happy at Applehurst Farm, and she didn't own him a nickel. She said thank-you and if he ever wanted them back, he was welcome to them. And that's the way it went. Phyllis ended up with Uncle Philip's cows and took pride in preserving the Brookvale bloodline.

Ironically, the herd was evicted from Applehurst – physically removed – on the same day as Uncle Philip's funeral. Stranger still is the fact that Phyllis sold the Brookvale cows to another of Philip's nieces, on Aunt Virginia's side: (Sheryle and husband Steve Hurd of Hampton, (who went out of business within a year).)

The morning following Philip's big wake, our uncle's casket was carried into the cemetery within sight of the Peterson farmhouse and, indeed, the very bedroom where he was born and died. Phyllis's dog Arnold ran loose, looking frantically at all the people, nervous. He peed on a gravestone as mourners tried not to notice.

It was a most unusual burial, since Philip had requested cremation, yet he was still there in body with several hundred well-wishers standing in the cemetery and along Martin Road. Uncle Philip's body traveled on his 1942 Federal flatbed after leading an antique tractor parade touring the farm. Friends and neighbors drove the 20 tractors slowly as the crowd watched from the graveyard. The rambling procession took nearly an hour to meander its way back to the cemetery as a Channel 9-TV cameraman recorded the whole thing. John Brewitt, who grew up with us to become another of Epping's Brewitt undertakers, said he would wait until everyone left before loading the casket into the hearse for cremation back at his shop.

Lifelong family friend Charlie Kimball declared loudly from graveside: "You ain't nevuh gonna see another send off like this one! No-suh!"

As Philip's oldest niece, I considered saying something to honor Philip's passage and request prayers for the 'Brookvale bloodline' which was also passing that day. But there were plenty of others on the agenda with prepared speeches. I didn't step forward.

Much later, still in the cemetery with 'The Brewitt' waiting patiently for his chance to snatch the casket, Phyllis chatted with the Hampton cousin who took the cows. Sheryle Hurd said husband Steve had stayed home to keep an eye on "Dear Philip's" animals. She said they were pretty confused about how to fit in with the rest of the herd and didn't seem to understand fences.

"They escaped four times during the first hour," she said as Arnold peed again on a nearby shrub.

Phyllis's Friend Arnold

The way we were ...

46. FAMILY PLOT

Our own cemetery, the Harvey Family plot, is easily seen from Nottingham Square Road. It sits on a rise – or, more correctly, it IS a rise – on the east side of the two-lane asphalt. Same as our house, the burial ground was built up and back-filled because the ledgy ground is too solid. Digging holes six-feet deep would be impossible because of shallow bedrock. Our graveyard lies in the Night Pasture, directly across from 'The Edgerly Place', Phyllis's farm.

Unlike nearly all of the other 70 private cemeteries scattered throughout Epping, ours is located on land still owned by the family buried in it. During my lifetime, it has accepted five more bodies – my Harvey grandparents and two hired hands who we cared for through their physical demise, plus our Uncle Cos from Connecticut. Also, unlike most other small cemeteries, ours still sees daily aboveground action with family members tromping back and forth across it between the main farm house and Edgerlys' and a stump dump on the west side of the road.

Whoever originally laid out the graveyard tried to set us on a more formal track than Dad has been inclined to follow. The oldest Harvey monuments are of white marble with matching arched tops, a dozen of them, two rows of six, all lined up straight and square on granite pedestals. Central to this six-pack theme is an eight-foot-tall obelisk tagged with "Harvey." An intriguing carving adorns this marble marker. It is of an unidentified man in a mourning cloak resting his face on a handkerchief while leaning on a mini-likeness of the Harvey obelisk. A poem about love and grief is finely etched beneath the carving, but we can't read it anymore because acid rain has dissolved the script along with part of the man's rear-end.

When I was a kid, granite slabs were stacked as a four-foot-tall retaining wall to enclose approximately 2,000 square feet of burial space. Cast-iron posts were drilled into the wall and draped with twin chains to further define the perimeter. Roughhewn stone steps still face Nottingham Square Road. My youngest sister Erica installed a granite post with a wrought iron hook and hangs baskets of flowers there all summer. It is good of her to do this. The rest of our cemetery wall and chain fence were leveled by Dad many years ago.

My father's extreme makeover of the family plot was another of his (some-would-say) 'desperate' responses to Epping's adoption of land use ordinances during the 1960s. He anticipated that the town would begin making rules about where to bury dead people, and he already KNEW where WE would lay our weary heads for the last time – out behind the barn. Nobody was gonna tell Daniel Webster Harvey where to bury the bodies! So, one day out of the blue, he began raving about a place where we could buy defective reinforced concrete products. You'd think he found the end of a rainbow in Massachusetts, he was so excited.

"Ron Cote found the place in Wrentham," he said, referring to an earth-moving contractor who had recently moved his business from below the border to Epping. "They sell cracked burial vaults ... and culverts, mostly. But most of the cracks don't go all the way through. We'll just have to look 'em over." He said he had reserved a dozen vaults for $50 each and he could help himself to crushed culverts for free. Cote was willing to haul them in his truck.

WHAT on Earth did we need burial vaults for?!! None of us could get our heads around that one. Did we REALLY need to expand the cemetery right now?

"We're all gonna go sometime," Dad said. "And I'll probably be the first to test drive one. Besides, we have to do it now – stake out our claim – before some knowin' puke tells us we can't."

Mom's disgusted "FORGET it!" often went unheeded. And this was one of those times. She asserted that there was "NO WAY!" she was going to be left to rot in a casket inside a water-logged concrete vault. She said she didn't even want a casket. "Just dig a hole and lay my body in it. Let the worms do their thing and be done with it, natural like." But that wasn't Dad's idea of a good time. He still expects the family to come together in the end, one way or another.

The following day, he began messing up the cemetery.

First, he used his Hough loader to shave earth away from the east-facing wall, the side least visible from Nottingham Square Road. He scoured a 40-foot-wide gouge back into the Night Pasture as 20 heifers watched for their chance to join him in the mud hole. And there WAS MUD in the hole under the water during that unseasonably wet summer when none of the hay made it into the barn without a soaking.

Dad dug down to ledge. He banged and scraped his mechanical shovel against the fissured bedrock to loosen chunks until he gave up. He could not make it deep enough to provide six feet of earth above his fascinating vaults. But that was okay. He would haul dirt and cover the chambers with that, build the burial sites up, same as his forebears.

It took two days of thrashing between downpours for Dad to set the scene just the way he thought it would work, with five loads of fill stockpiled in one corner. I can't recall anyone helping him. Probably no one did, since we all thought it was a dumb exercise and hard on the loader. Dad's continual crashing behind the barn annoyed us. Besides, Mom was against it. Being on the low end of the pasture, the graveyard expansion looked more like a new pond, a reservoir, perhaps a pumping station sanctioned by the Fire Department. Every time it rained, the hole overflowed with runoff water.

There were, of course, other things Dad could have been doing. But the cemetery caper was his top priority. It was a major setback when Ron the contractor said he wouldn't be able to fetch the vaults until the following week. Dad began fretting that his covert operation would be reported to town authorities. The excavation was not totally-visible from the road, but clear enough if someone was interested enough to crane their necks a bit.

"No tellin' who might come over the hill and look at just the right time," Dad agitated uncharacteristically. It wasn't that he was afraid of getting caught. He was used to being caught. He just wanted to be clever this time. Clever enough NOT to get caught. You see, there wasn't even a law prohibiting the expansion of family cemeteries – not at that time. It was all in Dad's head, this assumption that government would take total control of privately-owned property. There wasn't anything to catch him for.

"People shouldn't have to go, hat in hand, and ask permission every time they want to do something on their own land," Dad has said a thousand times.

This scene, the groveling before the recently established Planning Board, must've kept Dad awake many nights. He thought of himself as a "wise old owl," always asking: "Who do they think they are? Who in their right minds? Who needs that? Hey! If I can't do what I want on my land, who can? Who can do it better and why? 'Who?' is what the wise old owl says: Who? Who? Who?!"

Considering the hours he spent playing out the various scenarios, it was inevitable that he would move to expand the cemetery. Of course, he would. It had it all – the land use issues, the rebelliousness, the responsibility of heritage, our name on the rock. I can see that now. But, at the time, we all thought Dad was being impossible. None of us wanted to think about dying. We didn't want him to be thinking about it either. He wouldn't be leaving us for a long, long time. And neither would Mom. Certainly not ourselves. And leaky burial vaults! We didn't think so!

"It'll come to you later," my father said. "Someday you'll realize just how smart your Ole' Pappy is! You just wait." And before any of us had time to wait, Dad and Ron Cote made their early-morning dash across the state line and picked up twelve preselected not-too-bad burial vaults covered by a tightly-lashed blue tarp. Using the John Deere bucket tractor, and with a herd of Holsteins watching, Dad dragged Cote's flatbed through the muddy pasture, to and from the cemetery, leaving its shrouded load behind.

Then it resumed raining. A prolonged drencher that prevented Dad from positioning the vaults within the fresh excavation. The hole filled with water and surely would've floated the boxes.

Dad just had to wait until the water receded and the mud dried up. A week later, he got Leslie, the hired man, and Matthew to help him slide the vaults into the excavated depression. But they kept floating every time it rained. So, with one eye on the shifting concrete boxes and the other alert for the anticipated inspector or the nosey neighbor who might ask just what the heck he was up to this time, Dad spent a wary two weeks.

Beside the still waters of the farm pond.

Harvey cemetery engraved obelisk.Albert, Phyllis, Robert Kimball

The Cub in happier times. 'Hanger North' aka '12th Wonder of World' was built by Dad because he had acquired a huge log & free trusses. Took years recruiting help to erect it: Matthew, Danny, Steve, Tom, etc.

Part Five – Farming Flame Gutters & Dies

The way we were ...

47. DEMISE OF SADIE LEE

Note: This happened around 1964 – same as the cemetery vault incident.

If 'Sadie Lee' didn't die, we could have displayed her in the Rochester Fair freak tent as the "Luckiest Cow In The World." But she wasn't that lucky.

The five-year-old milker was just one of several cows to be hit by lightening in the 20 years before the herd moved into a covered feed lot. None of us kids ever saw the lightning actually strike. And neither did Dad. But we accepted his explanation because, before I was born, he had found a cow burned and dead under an apple tree. He said cows often stood under trees when it rained. Then, sometimes, the tree would be hit by lightning which either came down the trunk or connected back up through the roots. If it didn't kill them outright, the electrical charge seared hair and flesh off any cow that got in the way.

"If you're ever in the field and lightning begins to strike, go sit in a truck if you can find one," Dad said. "Whatever you do, DON'T stand under a tall tree. Or a short apple tree!" He has always had this thing against apple trees. And snakes. Just doesn't like them.

Since injury by lightning strike is not readily overcome and scars last forever, the herd always had one or two walking survivors with hairless blazes along their backs, rumps and usually down one rear leg. We gave these unfortunates names like 'Scabby Back', 'Blister Butt' and 'Scarlett' (not to be confused with 'Bag Dragger' and 'Tickle Breeches'). They usually recovered with time – shell-shocked and dazed for a few days following the hit, a bit slack in their appetites – but eventually got back into the milk-yielding routine.

Sadie Lee was a favored cow because she had been such a pretty calf. Part Ayrshire, her color was mostly reddish-brown mahogany, instead of Holstein black. As a calf, she had the cutest little speckled nose and symmetrical white patches across her shoulders and on all four legs. She must've had Jersey in there, too, because she wasn't very tall and her head was blocky with a short nose, dished face, and fluffy ears. A coppery cow-lick curled across her forehead. Sadie was very sweet. Many a cud was chewed within her chubby little jowl. Then, because she was a good-natured heifer, Dad let her keep her horns, didn't chop them off with the other yearlings. Sadie's horns curved gracefully inward and were seldom used against her herdmates.

It was a wild summer, that year when our pet was hit by lightning, the same summer Dad expanded the cemetery. Each day, the air remained heavy with humidity as a glaring sun warmed temperatures into the nineties every afternoon. Then, almost without fail, high cumulous clouds began rumbling in the west, then closer and closer with increasing intensity until we found ourselves – family, friends, hired help – gathered in lax little groups, seeking shelter from full-on downpours, thunder, lightning and sometimes hail.

During these weather events, the best place to be was sprawled among bales of hay in the good company of the (probably wet) hay crew. All of our growing-up summers were spent haying, so we seldom missed thunder storms while working. Sheltered within the several old post-and-beam barns, we stayed back about 20 feet from the gaping doors. Orange-breasted swallows returned to their nests and swooped overhead while we watched the storms through the doorway. Often, mist from wind-blown rain covered us, yet we felt safe inside the creaking beam-and-barn-board structures.

That sense of safety was not enjoyed inside the farmhouse during electrical storms. Mom always told us not to leave any lights on. So it was dark indoors and very noisy with the wind blowing and window glass rattling with each ferocious crash. Sometimes the phone tinkled and the electrical outlets snapped continually.

So, if we happened to be at the house when a storm hit, it was better to be outside. As many as a dozen of us would sit on the porch, open to the south, facing the driveway. From this vantage point, I have seen the barn hit twice and the mulberry tree zapped once – not all in the same storm. Our artesian well pump has been knocked out multiple times. Both the tree and well are located in the middle of the yard with the barn just to the right. So, as storm water washed deep gullies straight down the driveway from a trench silo Up Back and lightning gave us pause, the porch always felt like front-row center in a big-screen thriller. Afterward, we claimed to smell ozone.

"Ozone from the lightning," we'd say. "Smell it?" This was back before anyone mentioned holes in the stratosphere.

One afternoon following a real crasher, Sadie Lee came staggering across the pasture toward the farmhouse. She had been hit. Her back was a mass of cauterized blisters and she smelled of singed hair. Even worse, though, was Sadie's condition as an expectant mother. She was due to deliver her third calf in August, two weeks after the storm, during that horribly rainy summer of 1964.

Whoever was around (I can't remember who) helped Mom direct Sadie into one of our two 'comfort stalls' on the sunny side of the barn, where, hopefully, she would recover. Our family veterinarian, Doctor Ramaika, was summoned from Dover. He said he'd come first thing next morning. In the meantime, he said to put molasses into her water and see how much we could get her to drink. The cow drank all right, but remained restless.

When he arrived next morning, Dad told "Ramaika" (that's what we called him – "Doctor" to his face). Dad told Ramaika he had checked Sadie twice during the night and, as near as he could tell, she hadn't laid down or slept at all. The large large-animal veterinarian did his usual thing. He stuck a needle into the cow's neck artery and drained a bottle of electrolytes into her. Then he injected a shot of something "to calm her nerves" and sprinkled the wound with sulfur powder. He said there wasn't much else we could do, except encourage her to eat.

"We'll just have to wait and see how she comes 'round," said the vet.

Two days later, Sadie Lee was still drinking her molasses-flavored water, but hadn't eaten or slept much. Indeed, she appeared to be in labor. Her udder was engorged and her low-hanging belly bulged with the calf she carried. She fretfully scuffed a big circle into the deep sawdust bedding inside the box stall.

Dad called Ramaika again.

His late-afternoon arrival was the beginning of a very bad ending for Sadie Lee. Her chances for survival grew slimmer with each passing half hour. First, the veterinarian confirmed that our favorite blister-backed bovine was in labor. Then, he announced that her calf was dead, stillborn within her womb.

"Feels like a big calf, probably a bull. Could've been dead a while, probably since the lightning," said Ramaika after his initial vaginal examination. He looked worried and explained that the calf would have to be extracted "in pieces" because – though the body was not stiff like when rigor mortis sets in – it was difficult to work with a lifeless carcass within such a confined space. Today, Caesarean sections are common in addressing complicated livestock. But forty years ago, it was not an option. Our veterinarian said he had a flexible wire saw and promised to do his best to remove the bull calf without hurting our already-damaged cow.

It's really hard to watch a procedure like the one Ramaika undertook, so Susie, Matthew, Phyllis and I went about our chores and left Dad and our hired man, Leslie, to do the dirty work of assisting the vet. Truth is, our chores had already been completed before the familiar blue station wagon pulled into the driveway. We just wanted to know, from a distance, whether or not Sadie would be okay. Her occasional long low moans gave us reason to doubt.

I'll always remember the sweaty veterinarian's words, spoken long after fetching his surgical equipment and working with Dad and Leslie holding Sadie upright at first, and then lying her down.

"Damn!" he exclaimed, and then repeated: "Damn! I'm sorry!" He said it wouldn't work, that the unborn bull was somehow twisted around inside the mother and he couldn't get a good feel for where to begin sawing with the wire. "The two places I've tried are mixed up with her uterus and if I cut, I know I'll hit her, too." She would quickly bleed to death.

We kids stood silently as the men mumbled about what to do. I remember kneeling with Susie and weeping as we stroked Sadie Lee's sweaty throat and scratched her favorite spot, under her topknot. Dad headed for the house to call our neighbor 'Roland The Butcher', father of 'Little Roley'.

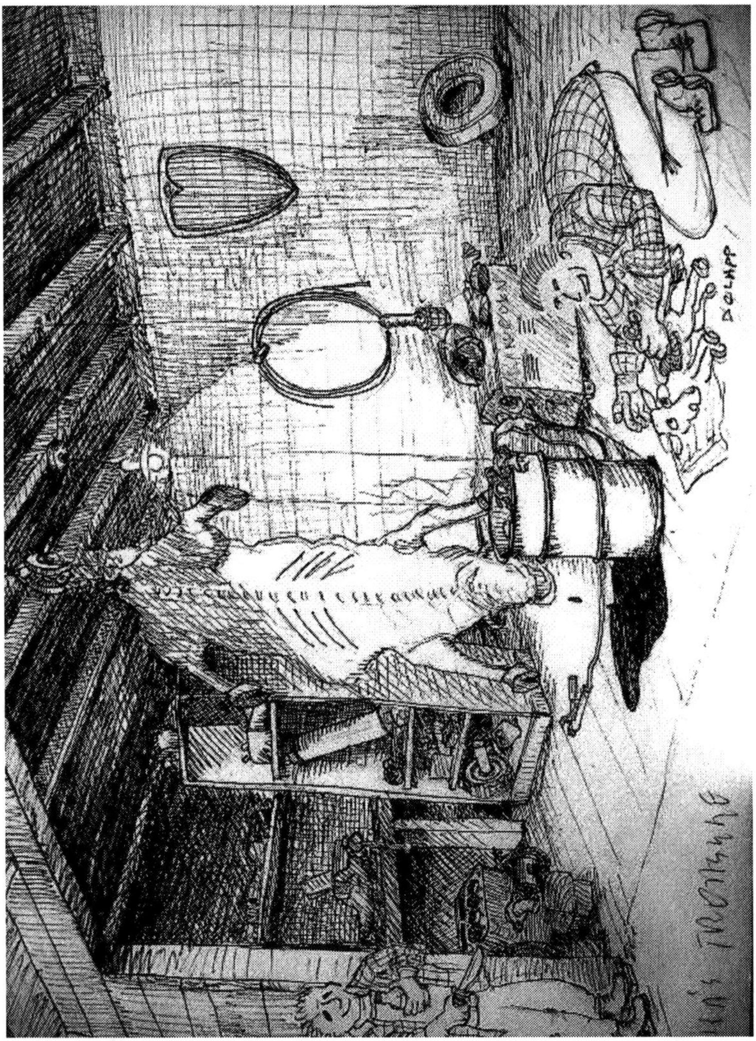

The way we are ...

48. HARD LANDINGS

A week before we left Epping for New Zealand this year, Phyllis's Piper Cub crashed into Red Oak Hill at the most unnerving angle. Phyllis, police, paramedics and others had dealt with the situation by the time David and I arrived an hour after the crash. A small crowd gathered to study the odd sight, clearly visible from the vegetable stand and road.

Walking in past the parents' house, we were approached by a young man in white shirt carrying a microphone.

"Did you see it happen?" he wanted to know and told us the pilot was someone named 'Paul' who had been taken to Exeter Hospital, not seriously injured. I think the kid had recognized me as 'a Harvey' because of my face. That's why he was wrong in thinking I might know something. He walked with us to Epping's police chief, Greg Dodge, who was in my sister Cindy's class, ten years behind me. Greg acknowledged that the kid himself had a well-known face, being a TV-Channel 9 reporter.

I thought: "My! Wouldn't this accidental sculpture make a fine tourist attraction to bring in vegetable customers! If only we could encase the mangled plane in a column of Lexan and let it stand to mark the end of an era." A sculpture entitled: 'That's IT!"

Approaching from downhill, the antique Cub's bright yellow tail and dislocated right wing formed a perfectly unnatural checkmark, as if a giant teacher had Okayed the field for landing. Or a big yellow exclamation mark against a deep blue sky. Viewing the scene from uphill, looking down, the broken flying machine reminded me of an exotic 'Bird of Paradise' in a floral arrangement. The colors were striking: sharp French's Mustard Yellow against soft autumn red and orange hues of hardwoods bordering the verdant hayfield, and, of course, the Cerulean sky reflected in the lower farm pond.

Yes. If it weren't such a disaster, the wreck COULD have been a work of art.

But Phyllis averted her eyes when the on-air reporter asked what it looked like to her. On the evening news, she declared: "I don't WANT to look at it!" She said it made her sad to see her "bird" like that, yet nearly laughed commenting on how she'd "have to say it appears that someone had a hard landing."

In New Zealand again, I received a letter from Phyllis saying that she was encouraging her friend Paul to reconstruct the 1939 Cub, though crash-site consensus was that it would be impossible to repair well enough to pass FAA inspection. Phyllis and Paul had hauled a twisted wing from the hilltop hanger to her Edgerly shop.

"The memories are the killer," she wrote. "Cleaning the shop to make room for the Cub, pawing through Michael's meticulous things." She said she had rarely entered the workshop since Michael left 15 years ago. Michael had been her partner for nearly a decade, but she finally asked him to leave and has never had as solid a relationship since.

Michael was her flying instructor. When he moved to Edgerlys', he worked with Phyllis continually. They cared for several horses and beef cattle, repaired the house and barn, and rebuilt the Cub (after a previous crash). They trained the horses to drive in harness, pulling a buggy they restored. I never understood what was bad enough about Michael to kick him out. They must've gotten on each other's nerves.

Toward the end, Michael worked part-time as an Epping police officer. He was an excellent marksman, good with guns. He also flew Leer jets and was on-call for local corporations to shuttle executives hither and yon. Once, he stopped David and me for speeding on Route 125. We could see in the mirror that his face was very red as he approached our car. "Oh gosh! This is embarrassing," he said and let us go.

After Michael left Epping, he rose through the ranks at SigArms, a local handgun manufacturer. After that, we didn't hear much more about him at all.

Not until ... last year.

My mother sent an e-mail reporting that Michael had "burned up in a fire" at his house in Nottingham. The damper had failed to regulate the wood stove during the night and all was lost. He was alone.

Within a week of Mom's news, we received a handwritten letter from Phyllis. It was about Michael's death. She said the police had phoned her first because, she guessed, Michael had listed her somewhere on an emergency contact card.

"They just left a voice message and it was very upsetting." She wrote about "making a speech" at the funeral and crying with Michael's mother and sisters.

More recently, Phyllis summed up the Year 2008:

"All of these funerals!" she wrote. "What a year! First Michael, then Ed (her milk truck driver who died in a Nottingham roll-over in July) and Arthur Scruton and Uncle Bill. And now Uncle Philip and the Cub! All that's good is that the cows escaped Susie's death threat and Paul wasn't hurt in the crash."

Paul's wife, Nancy, had been a passenger just minutes before the accident. She got out because she didn't like the way the engine was sputtering. "Smart woman!" wrote Phyllis. "There isn't much left of the front seat.

"All's we CAN do is keep pluggin' away. We lose our friends, seems like EVERY day and only The Lord knows why. Only HE can help us understand and cope with the daily routines by accepting HIS blessings whether they're joyful or sorrowful."

She wrote that working on the plane makes her think of Michael and 'pushing-on' with a 'we'll fix it' attitude. My sister's report sounded optimistic.

But the letter we received a few days later wasn't so upbeat. Apparently, Susie had enlarged and laminated photos of the crashed Cub:

"At supper the other night, Mommy pulled out these horrible blown-up huge pictures and expected us to use them as placemats. I turned mine over and quietly left. Why does anyone think these images are even interesting? Mommy pointed at one and said: 'Look! It shows that Fernalds had already chopped the corn.' Am I supposed to enjoy looking at something like THAT?! Where is the respect? The sensitivity?

The privacy? Why don't they just make placemats out of Uncle Philip in the casket and lay THEM out for supper?!"

She had me. Hadn't I wanted to make a statue?

"That's IT!"

Louise's Farm Stand. Chicken and horse barns have lacked both for many years.

Back to 'Square One' at ease with Niece Analesa.

The way we were ...

49. BACK TO THE PLOT

The sun had already set by the time Sadie Lee began her death march out through the feed alley to our machinery repair shed. The men decided to move her closer to the chain falls. That's what they call the hoisting contraption, a block-and-tackle rigged with chain for lifting engines and tractor rear-ends … and dead cows.

Amid an amber glow of 60-watt lightbulbs, the cow carried her head low. Her dry leathery nostrils took one long last sniff of stale feed left by her herdmates who watched respectfully from behind the slatted barrier. Our pet's back was agonizingly hunched upward as futile labor contractions continued and milk dripped from her teats. Sadie's stiff legs shuffled forward toward the spot outside the repair shop where Roland's son, Roley, promptly dispatched her with his .22 Magnum and buck knife.

As soon as we heard the shot, Doctor Ramaika rushed up the alley, leaving us behind sobbing. "You kids, keep your distance! Wait for your mother!" he ordered.

The veterinarian was beat. He said he wanted to do a quick autopsy and go home. Matthew had followed our cow up the manger, since Dad seemed to be missing. We imagined our brother and Roley hooking chains around the fallen milker's rear legs. That's when we realized it had begun pouring rain outside. Tin roofing roared overhead as a row of 60 sad-eyed cows watched our odd human behavior huddled in the concrete feed alley.

Mom appeared from the depths of the barn with Albert and Cindy trailing after her. She drew our sad little bunch together in a comforting hug. With tears in her own eyes, she delivered another of her explanations about life and death. Although I had heard it several times before, she told us about putting things out of their misery.

"It just isn't right to let an animal suffer," Mom said, assuring us that Sadie Lee had been a good cow, "a special cow" and we had done our best to give her a good life. "And now, with the lightning hitting her and all, she has suffered enough. We had to put her out of her misery."

Still, we wept while walking up the feed alley. We had seen slaughtered animals before – cows, pigs and chickens. Or I, at least, had seen them plenty. And probably Susie and Phyllis had, too. Certainly Matthew. But Albert and Cindy may have been too young. So we thought we knew what we'd see in the shop. Sadie's lifeless body would be dangling upside-down from the ceiling. And Little Roley was probably already slicing her in half, down her belly, pulling her skin off. He would be operating with one of his several butcher knives and, later, a bloody hack saw.

But, as we accidentally splashed into mud puddles while crossing the dark back yard, we were surprised to meet Little Roley standing in the shed doorway. For once, his hands were empty. No rifle or knife, no saw or even a beer bottle. Plus, he was uncharacteristically smiling from ear to ear.

"You will NOT believe this!" he exclaimed. "I can't even believe it. And I'm the one who cut her open!"

We didn't know what to think as we entered the mechanic/butcher theater. A single droplight illuminated the red-and-white hide of our dangling dead pet. A rusty barrel had been placed beside the carcass to receive her warm entrails.

All of this was normal.

So why was Roley outside? And smiling?

And what was Doctor Ramaika so excited about?

The big man was on his knees behind the gut barrel jabbering at Matthew. At first, I thought they were trying to revive the sleek wet bull calf that lay so obviously dead on the oil-stained concrete floor. But that wouldn't have been right!

"Come! Louise! Look at this. It's a miracle, I think!" crowed the exuberant veterinarian. "I didn't even know she was in there. And she's ALIVE!"

We rushed forward and WOW! The cutest little, mostly white, calf lay there in a jumble of wrinkled grain bags. Ramaika continued rubbing it vigorously with one of the sacks, probably just to stay in contact with his amazing find. The baby heifer wagged her wet head around, not yet aware of her surroundings and the nine sets of human eyes admiring her so lovingly. Our little sister Cindy scrambled to stroke her and squealed joyfully.

"Look! She's red like Sadie," she said.

"That's because she's an Ayrshire," Mom said.

After making sure the awesome heifer would be all right, Doctor Ramaika said he was done for the night.

"I'll give you a call tomorrow. There's something I need to explain, but not tonight. It's been quite a day!" he declared and disappeared.

Little Roley accompanied the vet back through the barns to the front yard where our butcher retrieved a six-pack of Budweiser from his truck. The legal drinking age was 21. Roley was a 17-year-old high school dropout, usually armed, at least with his buck knife which he kept under his pillow at night. Never far from a rifle, Roley was always on call to shoot dogs, cats, woodchucks and skunks, besides, of course, killing animals in the family slaughter house. It was okay with us if Roley drank. Who would question it while he carried a gun? Especially since he was using it to do our dirty work.

Little Roley was expertly tugging the hide off our dead cow when Paul McPhee's two young nephews, Mark and Tommy, stepped into the partly-lit shed.

"Yowsa! She's wet out there!" Tommy blurted looking half possessed with his bulging McPhee eyes. The brothers slapped water from their ballcaps. The shop resonated like a drum from rain pelting round-edged boards along the west wall. "Ya seen Uncle Paul?" Tommy asked.

"Shit! What're ya doin', Roley?" asked Mark. "Can ya spare a beer? We'll split it." The two boys were still in high school because their parents wouldn't hear of their quitting. But much of what they learned – things like hollering, drinking and swearing – was picked up outside the classroom. They had joined the McPhee "Wild Bunch" and were good friends with Roley. Their mother, Lucy, is Paul's older sister and married to an evangelical minister with a church in Exeter. Needless to say, The Reverend and Lucy weren't always chuffed with their sons' attraction to The Good Life on Uncle Paul's farm.

Roley gave them a nod. "Take two. They're small!" he laughed. "We're celebratin' the arrival of that little heifer."

No. We told them we hadn't seen Paul.

"But, now that you mention it, where's Dad?" The guys laughed at my question because they had all been through this before. Dad could be anywhere. But not HERE. Not where an animal had been killed.

That's when it came to me, this vision of life continuing through the next generation. Sadie died; her calf lived. And I knew, for certain then and there, that Dad was in the cemetery. I slipped out into the blustery night to find him.

On my way through the hay barn, I reached around behind a post to fetch a stashed flashlight. If I hadn't been so sure I'd find Dad on the other side of the junkyard, it would've been too scary to set off into darkness behind the shop, even with a flashlight. The junkyard was also cemetery of sorts, the final resting place for every vehicle that ever rolled across the farm. At night and in the rain, no rabbits or woodchucks would startle me, but the rusting relics cast plenty of shadows to make me jumpy … especially when metal creaked for no apparent reason.

I played my light across the crest of the Night Pasture knoll and illuminated the corner where my ancestors lay. Sure enough! There was Dad lit up like a ghost. He cast a giant shadow into the trees beyond the fence. Though I couldn't see very well, I knew he was standing on a vault in the middle of his unfinished excavation site. With my beam on him, I chuckled at the way he appeared to hover above the graveyard. Anyone who didn't know about the floating vaults would have been totally freaked out.

"Hey Daddy!" I yelled. "What're you doin' out here in the rain?" I was eager to tell him about the heifer calf, but it would be better if he saw her himself.

"That you, Daughter?" he called. Leaning on a pole, he explained that he was "orienting" the family by pushing the vaults around to straighten them. With water still in the hole, he had been fiddling with the new vaults all week, hoping they would settle down square to each other when the water receded. Then, maybe he could finally finish the cemetery pre-need expansion.

I had to hand it to my father! How many other Epping men would be like this, looking like Huckleberry Finn slipping away on a raft, floating across a cemetery in the middle of the night? Dad was "just orienting the family" in the dark, and in the rain, because he didn't want to deal with death in the shed. It made perfect sense to me. And to him.

But the general public would never understand.

'Did you hear the one about …?' Dudley and Daniel

The way we are ...

50. TORA BORA CAVE

We are now in November of 2008, still basking in the glow of Barack Obama's election to the U.S. presidency. Before leaving for New Zealand we voted absentee and didn't hear the outcome until reaching our "Cosy Dell" winter retreat. Despite missing celebrations with family and friends back in New Hampshire, it's not all bad being away to witness joyful international reactions to this year's American election. Totally opposite from bewilderment following the election of George W. Bush in 2004.

Kiwi friends phoned and e-mailed as soon as the Obama news broke. "Waahoo!" was heard on our message machine. People on the footpath and in the shops who were friendly enough before the election came gushing toward us with congratulations. Suddenly, everyone is feeling more positive about the United States. Before this, few Kiwis mentioned our being Americans. It seemed they didn't want to hurt our feelings.

One expatriate friend who has lived here for ten years said: "It almost makes me want to go back!"

We've received a letter from Phyllis – written before the election. She mentioned campaign rhetoric about terrorism, specifically Obama's criticism of his Republican opponent: "John McCain says he will follow Osama bin Laden to the gates of hell, but he won't even follow him to the cave where he lives!" Phyllis joked that Dad should be recruited for the 'War on Terror' because Dan Harvey has no qualms about pursuit.

"Daddy still comes over here every morning to read the paper and watch me do chores," my sister wrote. "Even Republicans had respect for the Tora Bora caves and gave bin Laden some privacy! Why can't I get the same? Things have changed, but Daddy thinks he can still question my intelligence and tell me I have 'no business sense', even though I'm outta there.

He criticizes me for wasting hay and asks when am I going to fix this or that?"

She suggested that with a model like that, it was no wonder The Sisters "think it's okay to attack our self esteem and invade our privacy." She asked: "Why do they always have to push, push, push, testing to see how much crap we'll take? It makes me tired."

I wasn't surprised by Phyllis's report, since the last words from our mother's lips – spoken to me before our six-month absence – were:

"Now, we just have to straighten out Phyllis!" I KNEW Dad would be the frontline spy toward that end. It's his habit, this straightening-out of Phyllis. Dad has been sitting around in the milkroom reading obituaries and watching Phyllis work with the cows for the past 14 years. He has always been full of suggestions.

Indeed, that is one of my family's least endearing practices – the offering of unsolicited advice.

I remember the project I was working on when I wised-up to the compulsiveness of Harvey advice. I was high-school-age – many years ago – building calf pens of my own design. As I worked, I anticipated that there would be suggestions … and, inevitably, I would resent them. But, for the first time, I realized that I was more aware of the dynamic. And I'd be better off – if only I could manage my reaction, my own bad habit.

These calf pens were unusual with raised-floors, self-cleaning, unlike any others. They would each hold two calves. Six pens made of 2x4s and 4-inch-wide pine boards from the sawmill. I designed, sawed and hammered the comfort stalls together, knowing they would work well and my carpentry was decent.

I gave the project quality thought and was prepared for criticism: "Don't let them get to you! You're doing a good job. Don't take it personally. It's just the way we are." And it was (and is) ALL Harveys' peculiar habit to offer advice, even if advisees are not prepared to receive it.

Not until later, when I began working in offices, did I learn that it is NOT a co-worker's duty to tell others how they should be doing their jobs. It isn't even a supervisor's right to continually critique performance. There are professional lines in the sand, not okay to cross. 'Harassment' was such a discovery for me! I like that about working off the farm, sharing workplace courtesy and getting paid to do a job. That's not my family's way, where we feel entitled to remain on each other's cases, always testing our mettle.

Accepting criticism has always been difficult for me. It's my one flaw. That, and failure to squelch an occasional cackle.

Phyllis wrote that Dad told her – (this was before the election) – that he intended to vote for John McCain, but don't tell anyone.

"Now, why would he do something like that? Vote for a Republican when he KNOWS every other member of his family is for Obama? Except probably The Petersons!"

She concluded, however, that Dad was probably "just yanking (her) chain and oblivious to how bad I feel being evicted by my own family. Why does he have to bother me with Obama, too?"

Phyllis wrote that she was starting a rumor that she planned to resume milking at the farm in January. She joked that nobody would notice in the dead of winter.

"Everyone just stays home and minds their own business after Christmas," she wrote. "When the heifers begin calving, I'll lead them over by the neck, one at a time, twice a day if I have to."

It didn't take long for word to spread. The next week, we received a letter from a niece, saying Grammy was worried that Phyllis would sneak cows back to the farm.

"But she's not going to, Analesa! She's NOT!" said Mom, as reported in her granddaughter's letter. Never a good sign when Mom speaks your name while delivering such a declaration!

Come January deep freeze, would Phyllis defy The Sisters' and our parents' cow-restraining order? Would she go ahead with her threat to reactivate the milking parlor?

Of course not! Not with Dad pursuing her to Tora Bora.

Phyllis wrote that Dad continues "interrogating" her every morning, asking about her plans and questioning her judgment and the ever-elusive 'business sense'.

"He kept asking me, EVERY morning: 'Did you get paid for the cows yet? Did you get the money in your hand?'" Phyllis wrote that she was reluctant to tell Dad anything about the painful herd dispersal. "... because why should it be any of his business, anymore? But he kept asking.

"So I finally said 'YES. I got paid! All right?' Then, next day, I got another 'love note' from Susie wanting more money for taxes. I thought it was between him and me. But NO! He tells me 'between you and me', he's voting for McCain. And then he runs and squeals on me to The Girls like a spy."

It is, understandably, difficult to break Phyllis and Dad's 14-year-long morning routine. My evicted sister stopped going to the farm every morning because she has no cows to milk.

And Dad ... Well, what's HE supposed to do? He doesn't seem to have considered the consequences of selling the cows.

For years, his routine was to fetch the paper from a box in front of the farm house, then go sit around watching Phyllis milk the cows, then deliver the paper to our sister Cindy's house where he visited with Paul McPhee, her partner of ... well ... 14 years. Then he drove a mile up the road to his friend Toots's house where he visited some more until it got to be about 9 o-clock and Madelyn, Toots's wife, had to go to work in the school cafeteria and Toots would drive her there.

BUT Toots died the same year that Cindy began kicking Paul McPhee out of her house. Toots couldn't help dying and nobody blames Cindy for separating from a man who refused to marry her. But it DID upset Dad's morning routine by not being able to go holler with his almost-son-in-law.

Dad probably still misses Paul's strong cup of coffee, which he continued accepting even after he was told to give up caffeine. He also ignored our cautions against submitting to Paul's Harvey-bashing harangues. I went over there a few times and witnessed the game of Dad's daily reports to Paul, who gleefully took what he was fed and spewed back unflattering observations on why our farm was going to hell.

Nothing new. I grew up with Paul. He has ridiculed us for as long as I can remember. It's his 'bad boy' image, his rogue appeal. The quandary is why did Dad go along with it, never defending us?

Now, with Paul McPhee's bitter coffee gone cold and Toots in the ground, what's an 88-year-old roamer to do before it's time to eat lunch and take a nap? Until recently, Dad made a point of walking down to the main farm for exercise. Now, his legs and back have become too painful. Still, he walked the distance for a good long while. Neighbors often stopped to talk as he leaned on his cane to 'take a breather' along the road between the parents' new house and the milking parlor.

With the cows gone, there is no longer a reason to walk or drive to the farm. And with his old house occupied by tenants, it's not as if Dad can just pop in to use the toilet. Phyllis is all that's left of the old routine. And even she's not all there, having retreated to her own farm.

Dad, like Phyllis, resists the forced break in the rhythm of farming. My sister threatens to return to the milking parlor in January while Dad declines to sever his tie with this daughter who he habitually double-crosses. He HAS to follow her to her cave.

It has always been this way.

I – being the most normal person I know – am puzzled by this father-daughter dynamic. Phyllis has always treasured Dad's role in her life, effusing about how "blessed" she feels every morning to have his company. She has told me how grateful she is that Dad is able to "share in the enjoyment of farming." Yet, Dad 'The King' has helped yank the rug out from under the cows.

No doubt, Phyllis was just being a brat when she threatened to return to the farm. Like Dad claiming to vote for a Republican.

She didn't. And maybe he didn't either.

Pictured: Otis preparing brick house addition for in-floor heat. Me with hibiscus in front of barn which will someday have greenhouse off back. Foundation (behind barn) has been in since 1971; never agreed on what to build on it.

Christa and Johnathan Weaver, Sam and Beth's kids.

The way we are ...

51. SETTLE OUR HASH

UPDATE EPPING 6/2010: I may leave this chapter out because Erica says I have been "very insensitive" in writing about our Harvey insensitivity, plus I "don't know anything" because I've been away. True. All true. She's right, of course. Still, it's my sad story and I can't write it any other way. This, after I've printed and given away 40 copies to family and beloved pre-readers. Fell short in distribution; now printing 100 more. I have a lifetime to get rid of them. The Sisters' new "Milk Room" soft-serve ice cream and doughnut emporium has become a great success. Lines of headlights coming from three directions converge on Red Oak Hill every night. Mom, Erica, Cindy and our almost-sister tenant Bonnie are making it happen. They have put two life-size plastic cows in the old milking parlor. One doesn't have a head, but nobody can tell since it is behind the faux herdmate, pretending to eat from the feed hopper. The plastic cows are certainly more sanitary and better behaved than real ones! They require less maintenance. Phyllis's remnant heifers are calving. She milks two at the Edgerly Place and claims to enjoy my writing. She says the book gives her "belly laughs." A brother called it: "kickass accurate." Another sister sent her copy to a TV talk show. People want more books. This is all good news (for me, at least). I guess I won't ax this chapter. The major factor in my decision is Niece Christa's sincere thank-you to me for writing the story and including her. "Five times!" she said her picture is in there. She told me this two weeks ago before her Senior Prom, which she attended and then flew to Homer the next day. As an Epping High School graduate, she is off for a summer of Alaskan adventure. HURRAH! FOR CHRISTA!!! I say in dedicating this chapter to HER:

We Harveys used to be happy with our wholesome little lives taking care of animals, harvesting crops and being good stewards of the land. People said we had it lucky and suggested we were decent and hard working. Hundreds of people came whenever we threw parties on top of the hill. We thought we knew how to have fun.

But now. Well, let's face it.

We're missing the point. I keep wondering if we lost our senses of humor overnight or if it was a gradual process.

"Christa won't help us clean her bedroom," Susie complained during a recent purposeful visit to my house. My sister explained that she, Erica and Cindy were working inside our former-brother-in-law Sam's house (also the home of our nephew John, 19, and niece Christa, 16 – when I wrote this).

"Cleaning up" was the term Susie used. The object of their cleansing was a house that The Sisters decided should remain in Sam's name when Beth and he divorced. They moved Beth to a nearby trailer. And the kids were expected to wander between the two.

Susie said she had asked Christa to take a garbage bucket out from under the kitchen sink.

"And she just left! Got on her bike and rode away!" My blond sister said she didn't think it was "right for these kids to go off to play" when their rooms are a mess. She wanted me to withdraw our invitation to take Christa to our camp in Maine for the weekend.

To me, it was understandable that Christa DID NOT and HAS NOT appreciated The Sisters' repeated unannounced intrusions to 'clean up' several of our siblings' homes – most aggressively Beth's and now her ex-husband Sam's residence.

I didn't know how to respond. So I listened as Susie described Christa's underwear as "a foot deep all over the floor." She asked if I would ever let MY kids be such slobs. She said, "We're making piles and taking half to Goodwill." She said Christa has a "thing" about "keeping every T-shirt she's ever had since kindergarten." My sister was irate about the situation, mad enough to pay me a visit.

I didn't want to say the wrong thing. But I could feel my collar warming. It was obvious to me that Susie had crawled out onto a limb that I might be obliged to chop off.

Was she really standing there assuming I had ever supported these home invasions? I did not intend to be her enforcer. Was she oblivious to my out-onto-the-wrong-limb attitude? Didn't she realize it irked me the way she has no qualms about ordering the nieces and nephews around and calling them worse names than 'dubbers' and 'dipshits'? It was I who told Otis and Amos not to take my sisters seriously.

"You've gotta make your own way," I told them, years ago. "If you start jumping every time they say to, you'll be jumping all your life and, one day, you will need to stop and make a big deal of it."

Cautious, I wondered how to avoid challenging my sister. Harveys don't do well with confrontation.

I stammered: "Well. Maybe it's not the garbage bucket ... Maybe Christa, you know ... Maybe she just doesn't appreciate ... you know ... you and Cindy and Erica going into the house and ... well ... going through all her stuff."

"Of COURSE she doesn't appreciate it!" blurted this sister who is charged with settling the farm estate when our parents die.

Her indignant response threw me. Here she was expecting to be appreciated for 'cleaning up' Sam's house. She saw nothing wrong with sorting through Christa's underwear. Or commenting on nephew John's personal belongings. She actually expected to be appreciated for her efforts! AND she expected to tell me who I could invite to our lakeside cottage.

"She's more likely to settle our hash than our estate. Lord, help us all!" I prayed as her truck churned gravel up the hill to exit our property.

The way we were ...

52. SADIE LEE LIVES ON

Mom knew about the freemartin problem before the veterinarian phoned to explain. And, on her own, she had decided what to do about it.

"Ya ... Ya, Ya. I thought so," she was talking to Dr. Ramaika when Susie, Phyllis and I came in from 'chores' the morning after Sadie Lee was slaughtered.

"Uh-huh ... uh-huh.." She listened as Ramaika told her that the miraculous little heifer calf could never be bred and couldn't produce milk because her blood had been infused with male hormones from her stillborn twin brother. The heifer was beyond sterile. She would grow up thinking she was a bull.

"A freemartin. Right. We had one before when I was a kid. Pa shipped it," Mom said into the phone. She understood the implications as we sisters eavesdropped in disbelief. But toward the end of the call, Mom was smiling. She kept saying, "Ya, ya" and "uh-huh," but her expression had turned puckish.

We were thrilled when she slapped the receiver down and declared: "We'll raise her!"

Cindy never knew how close she had come to losing the calf she had so immediately claimed as her own. She had been asking all morning what to name the little Ayrshire.

"Trixie! How 'bout Trixie? Or Surprise? Or Sundance? Jerry Lee would be good, don't ya think? Like Sadie Lee, but more like Jerry Lee Lewis" (who we saw on the Ed Sullivan Show). Our 7-year-old sister wouldn't shut up about it during chores. By suppertime, after oblique discussions concerning hormones and male and female characteristics and the bovine reproductive system, Cindy, (with Susie and me advising) decided to go with 'Jerry Lee' because a unisex name like that would let our hormonally-challenged freemartin swing whichever way felt right.

"I'll help you raise her like an ox," Mom told Cindy. "We'll all help with Jerry Lee, won't we?" She made Susie, Phyllis and me agree to help our little sister train the calf as a draft animal. Indeed, it would be raised as a steer, obedient to voice commands while pulling a scoot, wagon or sleigh. Raising Jerry Lee became an education for all of us.

The following summer, Cindy and her freemartin were banned from Mom's vegetable garden. The young ox couldn't be trusted not to drag its training scoot across the rows of carrots, onions and broccoli.

"That's ENOUGH!" Mom shouted early in the training. "Go practice across the road. See if you can get a track going around the pond." Cindy had her hands full with the frisky heifer beast. She worked with it just a little every day and was encouraged by Uncle Philip who had learned about oxen from his father, our 'Grampy Peterson'. Philip had raised two steers, "Rex" and "Teddy," as draft animals and was able to equip Cindy with small-to-larger yokes as Jerry Lee grew.

By the following summer, it was obvious that little Jerry had not beaten the 90-to-one odds that favored her being a freemartin. She was definitely a tomboy ox-type critter with male hormones eradicating what we, as 4-H cow judges, used to call 'dairy character'. Cindy's ox was neither sharp over her withers nor refined in her veining. She would have no need for a high, wide rear udder attachment since she would never develop usable teats. Instead, she was blocky over her shoulders and thick in the neck. She had big feet, a fat rump, and a curly coat like that of a beef animal.

Though she wasn't nearly as sexually alert as a bull and not at all aggressive, Jerry Lee was inclined to mount the heifers she ran with. Uncle Philip said we could probably train that urge out of her. He suggested penning the freemartin and a couple cows handy to the front yard so everyone could keep an eye on them.

"It shouldn't take more than a week to beat the urge out of her," said our uncle. "If you see her jumping on the cows, rush and holler 'Baaaah!' Threaten to hit her over the head with a stick." He said it was just an "experiment," since we didn't know anyone with freemartin experience. But Philip thought it might work. "She's just confused. We can probably straighten her out," he said.

After a week of being Baaaah-ed at and whacked, the freemartin seemed to repress its longing to become a father. We, at least, lost interest in monitoring her behavior in this regard.

Cindy's rapport with the ox was built on mutual respect and trust. Jerry Lee's robust strength and youthful energy were forces to be reckoned with. Sometimes she kicked up her heels and just ran away. Cindy had to teach her pet to "Giddup!" and "Whoa!" When the freemartin was small, our little sister walked around as if she had a dog on a leash, rather than a calf on a rope. Cindy tapped the beast's fluffy red ears with a cane to teach her to turn: "Gee!" for right; "Haw!" for left. Developing a language that the ox could understand was an ongoing project that showed all of us how very patient and determined Cindy is. She kept at it and celebrated even the smallest signs of progress. She growled at Jerry Lee whenever she did something wrong and reinforced positive responses by feeding her Gummy Bears. Over time, there was no need for a halter or rope and the cane became a mere fashion statement. The two of them became very good at responding to each other's body language.

It's a beautiful sight: a fit blonde girl sporting a blue-and-white plaid shirt and hook-handled drover's cane while walking beside an alert yoked ox obediently dragging a log from the forest. So much more romantic than an exhaust-belching skidder!

That's when people began losing their heritage of understanding the basics of agriculture – when technology like skidders came along to replace horses in the woods. As the timber industry and agriculture became dependent on gas-powered equipment, the government decreed that all milk would be pasteurized and vegetables should be wrapped in plastic. Horses and oxen were no longer safe to operate. Fresh milk and produce were no longer clean enough for consumption. Dairy herds disappeared from the landscape, relegated to concrete feedlots. Food production became a mystery.

I could tell it was happening by sniffing my classmates. In First Grade, John, Dawn, Rusty, Frances, Kenny and I were apt to smell of barnyard as we sat at our desks. Or, we'd notice it on our jackets when we went out for recess. But, by the time we hit high school, just Dawn and I smelled of morning chores. The other families – all of them more 'downtown' than Dawn's and mine – had signed up for Clover Leaf milk delivery, instead of keeping a cow. Dawn and I hated smelling that way. We called it our "social stinkma." But she had it worse than me, since her parents raised pigs, too, and they could reek.

The 1950s and '60s launched the "just imagining" phase of farming in the United States. Farmers were outnumbered 100 to one. People began "just imagining" what it would be like to live on a farm. You may remember Walt Disney and his Mickey Mouse Club productions. There at the beginning of television, the media master was alert enough to focus on the vanishing family farm. One series was called "Dairyland" with Annette Funicello being a movie-star milk maid. Disney's top Musketeer did her best to help kids imagine what a dairy farm is like. But I could tell she was just pretending. Her overalls and blouses were far too crisp.

Dad said the only part of agriculture the public understood was how to behave like sheep. He said the majority of Epping residents were happy to be driven around like a flock of sheep. "The dog barks and they all turn and run without knowing what they're afraid of. The planners 'just imagine' they know what's good for us. But, really, they don't know nuthin' about farming."

Of course, Dad didn't 'know nuthin'' either about how insignificant our farm was becoming within the scheme of American agriculture. Although we were never in the same league with Midwest farmers or large-scale operations in New York and Pennsylvania, we had a fleeting competitive advantage because we were close to Boston's urban sprawl. Plus, Dad had inherited the land.

But it was no use. Transport and refrigeration costs dropped for outside-of-New-England producers while ever-improving dairy technology flooded the market with surplus milk products (powdered solids, butter and cheese).

We were doomed. High tax assessments on woods and poor cropland prompted Rockingham and Strafford County farmers to sell for housing developments. Two local dairymen moved their families and cows to upper New York State where more fertile soils could support larger herds.

Dad continued to call himself a "farmer" and spoke up for dairy farmers – indeed, ALL farmers. What he failed to recognize was that we had become economically obsolete, outclassed. Progressive agribusiness producers in other parts of the United States (soon extending to Florida and California) didn't give two hoots about a two-bit 100-cow dairy in Epping, New Hampshire.

The USDA began seeing us as expendable. We were on our way out. Without federal price supports and a manipulated distribution system our dairy farm would've failed sooner. It must've been hard for Dad to depend on the government he disdained.

But Hey! No sense lamenting the past. We still managed to stagger into these modern times. And here we are with city people not understanding Harveys, and Harveys not understanding the basic appeal of materialism and economic influences that shape our lives.

Today's children are spared the intimacy and responsibility of 'chores'. They will never watch their grandfather comb the cows' tails on Sunday afternoons. Agriculture on a commercial scale functions on mechanized systems and equipment. Modern farm kids don't have a chance to hack frozen manure and ice from gutter cleaners and water troughs at 6 a.m., before walking to catch the school bus. Children are no longer splashed in the face with warm milk while training calves to drink from buckets. Instead, there are electronic control boards and calves are fed from multi-teat "calf-i-terias," adult operated only.

American cattle are presented with fodder delivered by feed mixer wagons. No need to wheel 25 barrows of corn silage. No need either for hayloft courage while dragging 15 bales of hay from one dim light bulb to the next as a blustery blizzard rattles the roof and pigeons coo from the vaulted darkness. No mutual pleasure from rubbing the underside of a heifer's outstretched neck. Today's cows have no names like 'Sadie Lee' and 'Judy'. They are just numbers displayed on large plastic tags riveted to their ears. They shy from petting, hoping for brief encounters with their human custodians who slash, rather than stroke, throats and appreciate udders for milk only.

So, even modern farm youth are disconnected from animal husbandry. They sit in the house watching television and playing video games, same as their in-town classmates who never did smell of silage. And, of course, United States dairy cows are denied access to pasture. They neither graze nor frolic. Their legs are stiff, their hoofs tender. They lack hope because it is a hopeless system.

Another casualty of The American Way. And, we, like a flock of sheep, have turned toward commercial progress. Out with the old! In with the new.

Jerry Lee, our freemartin ox has been dead 30 years now. May he/she Rest In Peace!

Epping "Annual Report, 2008" featured picture of our barn on cover (photographed before recent renovation). Also, around the same time, my parents received the New Hampshire Farm Bureau "Profile Award" for 'outstanding service to the state's agriculture and rural life,' including accomplishments through Grange, 4-H and town politics. The citation says Applehurst Farm includes about 1,500 acres, of which 300 are open and producing feed and vegetables.

And in the end ... EPILOGUE

WE'LL HAVE TO WAIT 'TIL THE COWS COME HOME.

Not sure who to quote. James? Or, Job?

"Resist the Devil and he will flee from you." (James 4)

"Thus far you shall come, but no farther and here shall your proud waves stop." (Job 38)

Or, Alfred Carr who said simply: "The jig's up, Boys!"

Ah, Alfred! How could you have escaped my musings?

He lived 33 years with us on the farm until contracting a miserable cancer that chewed through his sinus cavities and killed him in 1996. He arrived while I was still in high school and the 'Babies' Beth and Erica were still babies. An orphan 'ward of the state', Alfred had already lived on two other dairy farms before Aunt Edna Woodward arranged to have him move in with us. He had an official low IQ, but seemed as bright as the rest of us.

Alfred's other farms had gone out of business. He was very unhappy when he couldn't work with cows. That's why Edna brought him from Durham to Red Oak Hill. And then we let him down, too, with the herd termination program. Mom always said Alfred was "A Good Cow Man" and had that etched into his gravestone.

Alfred taught us things that only a wise man can teach. Lessons of love and humility, honesty and transparent integrity. He said the darndest things! The hired man's passing shone light on my own selfish disconnect from farm responsibilities. He died while we lived in Maine. Before that time, Alfred (who was about my mother's age) and I shared a special relationship of mutual support and amusement. He was always laughing at me.

Clued in from the start, Alfred didn't unpack his suitcase for the first two years that he lived in our attic. I ran after him several times pleading for him to turn around and come back. "I didn't mean to make you mad" ... "We'll be nicer to you, Alfred" ... "I'll buy you a new shovel!" ... "A hoe!" Once, he ignored my desperate offers and kept marching toward town, furious, a satchel in each hand. "I don't need you people!" he barked over his shoulder. I had to go get the car to retrieve him.

Alfred was no fool. He taught us to be more aware of personal boundaries. Unlike the rest of us, he had a temper and did not hide his frustrations. He drove Dad and my brothers to distraction, never hesitating to join in with seldom helpful observations, often about mechanics.

Once, Phyllis found him in her shop making secret Christmas presents. "Jeenus Man! There's sumthin' about it!" he declared as an unbalanced chair's legs became shorter and shorter while sawing to level the legs.

Each year, the artificial breeding service hosted a dinner at the Grange Hall. All area farmers showed up for the free food and a lecture. One year, Alfred, I, and others shared a table with the Scruton Family whose herd was large, all registered purebred Holsteins, many of them prizewinners, and their dairy operation progressive. Over beans and franks, the father, Frank, explained that they were milking the herd three times a day. (Twice is the norm.) We lesser farmers nodded politely somewhat awed by the concept ... until Alfred blurted out:

"The SAME cows?!! Jeenus Man! The SAME cows?!!" and the whole table choked with laughter, even the Scrutons.

A fountain of information, Alfred was sometimes mistaken for Dad. Mom kept a straight face, but laughed about it later whenever vegetable customers went on about how her husband had told them this or that, reflecting one of Alfred's rambling tales. I don't think she ever corrected them.

Alfred didn't have a driver's license, never drove a vehicle or strayed far from the farm, so it was especially entertaining whenever tourists stopped for directions. The goose chase began as they pulled out of the driveway. "You can't help but miss it!" Alfred enthusiastically called after them, waving.

Friends and neighbors gathered for Alfred's informal burial in our cemetery – McPhees, Allens, Fernalds, Aunts Ann, Edna and Ginny, Uncles Eddie, Bill and Philip. The previous day, Christmas afternoon, Matthew had dug the grave with his backhoe clawing through thick frost. He covered the hole deep in pine boughs so it wouldn't refreeze overnight.

During interment, Ron Cote's wife, Diane, testified that Alfred would be missed at the vegetable stand. "Once, all I needed was a few tomatoes," our neighbor said. "And all I had was a ten-dollar bill. So, I asked Alfred if he could make change. He said 'No'." I thought maybe he'd say I could pay later, but he didn't. So I asked: 'Well. What should I do?' Without skipping a beat, he said: 'I don't make change. You'll just have to buy more.'"

But, in the end, Alfred died. Susie's former classmate, John Brewitt who was destined for undertaking, supervised as the men handled heavy hemp rope to lower Alfred's casket into the gaping pine-fringed hole.

"I really like a good basic funeral like this one," said 'The Brewitt' as we took turns shoveling dirt onto the pine box. That's what we call undertakers in Epping: 'The Brewitt' because, for generations, Brewitts have done the job.

Alfred's demise was a harbinger of today. My youngest sister, the ever-giving Erica, stepped forward and assumed responsibility for our farm worker's hospice care. Always proactive regarding our parents' needs, she decided to spare Mom the burden and arranged with Phyllis to have Alfred live at Edgerlys'. Phyllis provided conversation, a roof, and open fireplace and food, while Erica ministered to Alfred's medical and personal maintenance needs. Toward the end, she had to help him eat. She took him to Boston where surgeons described his procedure as "The Big One."

Erica said Alfred had more patience and courage than the rest of us would ever demonstrate or understand.

I had made my bed in Maine – 2-1/2 hours' drive from Epping, 3-1/2 hours from Boston. I wasn't much help to Erica or the rest of them. Living apart from farm and family has its price to pay. Other family members supported Erica's efforts while I went to Boston only once. My heart was broken while looking into Alfred's sad bloodshot eyes, his mouth stuffed with wadding and sewed shut. I wasn't much help.

The story of my life: I'm not much help. I just think about it all the time – my family, the farm and my Epping hometown.

I've spent many years away. Yes. Now, nearly half of my adult life has been spent outside New Hampshire. It's not that I don't appreciate my heritage or love my family. I do. I have suffered no abuse. Epping has been good to me and made me proud of various accomplishments – personal and community. Can I help it if I've grown up to be a migrant, rather than static, farm worker? I feel better when I'm away from the family farm. Everything seems more complicated now and makes me feel small. No matter what I do, it's not enough. You know how that goes.

But just imagine!

What if EVERYONE – all of the "sheep," as Dad so unlovingly refers to us – what if we all thought about what's happening and changed our minds about the inevitability of growth and prosperity.

What if we looked at the McMansions and shopping malls that recently landed in Epping's every field, swamp and orchard and decided it's ENOUGH?

What if we sheep questioned authorities who projected in 1970 that Epping would become just like Manchester by 2020?

What if everyone looked at an aerial photo and saw that Applehurst Farm maintains the ONLY significant open space in town not cluttered with houses, swimming pools, Wal-Mart, Starbucks, Mattress Discounters, car sales lots, race tracks and horse paddocks?

What if we stood with conservationists and advocates of sustainable agriculture? What if we resisted the devils and they all ran away? Thus far and no farther!

You can't help but miss it.

What if Dad was right (Oh No!) when he said land should be taxed at its current use PERIOD!? Conversion penalties and impact fees do not slow development any more than an economic stimulus package changes a system that encourages greed. What if everyone suddenly saw what was right and did it?

You can't pay later. You just need to be more responsible now. The jig's up, Boys!

What if it's not just a quirky art decision that printed a glossy full-color picture of our c. 1813 barn on the cover of Epping's 2008 Annual Town Report? What if taxpayers begin appreciating their bicentennial farms? (Burley's have one, too.)

What if my nieces and nephews and our own sons (with their degrees from UNH, Harvard, Tufts, Universities of Maine and Pennsylvania) started an "Applehurst" farming school or agricultural daycare or summer camp?

What if everyone acknowledged and agreed to honor beautiful farmland, dense forests, healthy animals and fresh produce?

What if Mom's and Analesa's vegetable stands were sold out each day thanks to appreciative residents in the new housing developments?

What if we sheep demanded organic produce?

And milk. And soft-serve ice cream? Doughnuts, even!

What if Phyllis got her act together to be a source of organic milk, yogurt, butter and cheese, sold directly to neighbors from her own dairy bar?

Couldn't the cows come home then?

"I don't make change," Alfred said.
"You'll just have to buy more."

Burying Alfred Carr behind the barn. Identified: Cousin Dana Burtt, Ron Cote, Bob Kimball, Steve Martel, Tim Bruno, Sam Weaver, Brenda Lavigne, Toots & Madelyn Downing, Bruce Allen, John Brewitt and Matthew.